THE BRITISH ISLES

THE BRITISH ISLES

by

FREDERICK MORT, D.Sc., M.A., F.G.S.,
F.R.S.G.S.

Cambridge:
at the University Press
1914

CAMBRIDGE
UNIVERSITY PRESS

University Printing House, Cambridge CB2 8BS, United Kingdom

Published in the United States of America by Cambridge University Press, New York

Cambridge University Press is part of the University of Cambridge.

It furthers the University's mission by disseminating knowledge in the pursuit of education, learning and research at the highest international levels of excellence.

www.cambridge.org
Information on this title: www.cambridge.org/9781107632813

© Cambridge University Press 1914

This publication is in copyright. Subject to statutory exception and to the provisions of relevant collective licensing agreements, no reproduction of any part may take place without the written permission of Cambridge University Press.

First published 1914
First paperback edition 2013

A catalogue record for this publication is available from the British Library

ISBN 978-1-107-63281-3 Paperback

Cambridge University Press has no responsibility for the persistence or accuracy of URLs for external or third-party internet websites referred to in this publication, and does not guarantee that any content on such websites is, or will remain, accurate or appropriate.

PREFACE

THIS book should be used along with good orographical maps of the British Isles and the countries they comprise. Larger scale maps such as Bartholomew's Half Inch Touring Maps or the One Inch or Half Inch O.S. Maps will frequently be found useful, although a sketch-map is generally given in the text when reference to a large scale map is desirable. A fair general knowledge of the topography of the British Isles is assumed, and no space, therefore, is wasted on topographical lists.

A knowledge of names and positions is necessary, but is better obtained from a map than from a book.

The writer has consistently endeavoured to omit unrelated facts, and to select only those that illustrate geographical principles.

The book is suitable for pupils preparing for the Oxford or Cambridge Local Examinations or for the Higher Grade Leaving Certificate of the Scotch Education Department.

I am indebted to Mr John Deas, Headmaster of Belvidere Public School, for reading the proofs and for valuable criticisms.

F. M.

April, 1914

CONTENTS

CHAPTER		PAGE
I.	THE POSITION OF THE BRITISH ISLES	1
II.	THE BUILD OF THE BRITISH ISLES	9
III.	WEATHER AND CLIMATE. WINDS	15
IV.	WEATHER AND CLIMATE. RAINFALL AND TEMPERATURE	23
V.	AGRICULTURE AND FISHERIES	30
VI.	MINERALS	36
VII.	THE BUILD OF SCOTLAND	44
VIII.	THE RIVERS OF SCOTLAND	51
IX.	LAKES, LOCHS, AND ISLANDS	57
X.	VOLCANOES AND GLACIERS	65
XI.	THE HIGHLANDS AND THE NORTH-EASTERN LOWLANDS	70
XII.	THE SOUTHERN UPLANDS	77
XIII.	THE CENTRAL LOWLANDS	82
XIV.	THE CENTRAL LOWLANDS (*continued*)	89
XV.	THE CENTRAL LOWLANDS (*continued*)	93
XVI.	THE COUNTIES OF SCOTLAND	96
XVII.	THE BUILD OF ENGLAND	99
XVIII.	THE RIVERS OF ENGLAND	105
XIX.	LONDON	111
XX.	THE THAMES VALLEY	117

CONTENTS

CHAPTER		PAGE
XXI.	SOUTHERN ENGLAND	122
XXII.	THE SOUTH-WESTERN PENINSULA	129
XXIII.	THE SEVERN VALLEY AND THE BRISTOL CHANNEL	134
XXIV.	THE MIDLANDS	140
XXV.	EASTERN ENGLAND	146
XXVI.	THE LAKE DISTRICT AND THE PENNINE UPLANDS	153
XXVII.	NORTH-WEST ENGLAND	161
XXVIII.	NORTH-EAST ENGLAND	169
XXIX.	WALES	176
XXX.	RAILWAYS AND COUNTIES	184
XXXI.	THE STRUCTURE OF IRELAND	190
XXXII.	THE CLIMATE, AGRICULTURE, AND INDUSTRIES OF IRELAND	195
XXXIII.	ULSTER	201
XXXIV.	DUBLIN AND THE EAST OF IRELAND	206
XXXV.	SOUTHERN IRELAND	213
XXXVI.	WESTERN IRELAND AND THE CENTRAL PLAIN	217
XXXVII.	THE ISLE OF MAN AND THE CHANNEL ISLANDS	222

LIST OF ILLUSTRATIONS

FIG.		PAGE
1.	The British Isles and the Surrounding Seas . . .	3
2.	The Continental Shelf	4
3.	Section across the Continental Shelf and Scotland .	5
4.	Sketch-map of the British Isles showing the nature of the rocks	13
5.	Pressure Chart illustrating a Cyclone	16
6.	Pressure Chart illustrating an Anticyclone . . .	17
7.	Wind Rose showing the frequency of the winds from the eight principal points of the compass during January	20
8.	Wind Rose showing the frequency of the winds from the eight principal points of the compass during July .	20
9.	Wind Rose showing the frequency of the winds from the eight principal points of the compass throughout the year	20
10.	Tree showing effect of prevalent south-west winds . .	21
11.	Rainfall Map of the British Isles	25
12.	January Isotherms	27
13.	July Isotherms	27
14.	Gulf of Warmth over western Europe	29
15.	Diagram showing comparative areas under the chief crops in 1910	31
16.	Graphs showing areas under the chief grain crops from 1880 to 1910	33
17.	The chief coalfields of Britain	37
18.	Output (in millions of tons) of the three principal coal-producing countries of the world	39
19.	Graph showing coal exports	41
20.	Graphs showing production of pig-iron in U.K., U.S.A., and Germany	43
21.	A typical Highland Schist	46

LIST OF ILLUSTRATIONS

FIG.		PAGE
22.	Typical Highland Scenery	47
23.	The Lowther Hills with village of Leadhills in foreground	48
24.	Scenery of the Central Lowlands	49
25.	Cuchullin Hills, Skye	51
26.	Sketch-map to illustrate "piracy" by the River Avon	53
27.	Sketch-map to illustrate physical history of River Clyde	55
28.	Loch Katrine	57
29.	Loch Fyne, a typical Scottish fiord	59
30.	Sketch-map of Dumbartonshire Highlands to illustrate formation of islands and sea lochs	61
31.	Part of Island of Lewis showing remarkable number of small lakes	63
32.	Dumbarton Rock. A volcanic "neck"	66
33.	Rock surface near Loch Doon	67
34.	Loch Tay, an ice-eroded basin	68
35.	Map showing effect of ice-erosion	69
36.	View from summit of Ben Lawers	71
37.	Pass of Killiecrankie	76
38.	Routes across the Southern Uplands	81
39.	Section across the Central Lowlands	83
40.	Shipping on the Clyde	87
41.	Sketch-map of Central Lowlands	95
42.	Section across the Pennine Uplands	101
43.	Section from Wales to London	103
44.	Section from London to Beachy Head	105
45.	Sketch-map of part of Wey basin	109
46.	Map of London	113
47.	The Thames from the Tower Bridge	115
48.	Sketch-map of South-eastern England	119
49.	High Street, Oxford	121
50.	The Chalk-cliffs of Southern England	125
51.	Sketch-map of North Downs	127
52.	Sketch-map of the South-western Peninsula	129
53.	King Tor	131
54.	The Bore ascending the Parrett from Bristol Channel	137
55.	Typical limestone scenery	141
56.	Ruins of Dunwich Church	147
57.	Routes round Lincoln	151
58.	The Radial Drainage of the Lake District	153
59.	View of Derwentwater and the slopes of Skiddaw	155
60.	Dovedale, a valley in a limestone region	157

LIST OF ILLUSTRATIONS

FIG.		PAGE
61.	Railway Routes across the Pennines	159
62.	Sketch-map of the Manchester Ship Canal	163
63.	The Landing-Stage, Liverpool	165
64.	Carlisle as a railway centre	167
65.	Overlapping of the hinterlands of Liverpool and Hull	173
66.	The large towns of Northern and Midland England in relation to the Coalfields	175
67.	A distant view of Snowdon	177
68.	Carnarvon Castle	179
69.	Sketch-map of South Wales	181
70.	The Main Railway Lines of England	185
71.	Glendalough	193
72.	Graphs showing populations of Ireland and Scotland from 1845 to 1911	199
73.	The Congested Districts of Ireland	200
74.	The Giant's Causeway	203
75.	The Main Railway Lines of Ireland	207
76.	The Mallow Gap	211
77.	Rias of South-west Ireland	213
78.	Lower Lake, Killarney	215
79.	View near Sligo where the Central Plain meets the Western Mountains	221

The illustrations on pp. 27, 29, 103, 105, 173 and 181, are reproduced by permission of Messrs Oliver & Boyd from Dr Mort's *Regional Geography*, those on pp. 48, 49, 51, 57, 66, 68, 76, 87, 115 and 165, are from photographs by Messrs J. Valentine & Son, those on pp. 121, 125, 131, 177 and 203, by Messrs Frith & Co., those on pp. 46, 47 and 67 are from photographs kindly supplied by Mr J. W. Reoch, those on pp. 193 and 215 are from photographs by Mr W. A. Green, and those on pp. 71, 147, 155, 157, 179 and 221, are from photographs by Messrs W. L. Howie, T. Dexter, Abraham & Sons, R. Keene, J. Wickens and R. Welch, respectively.

CHAPTER I

THE POSITION OF THE BRITISH ISLES

THE British Isles are situated off the extreme western coasts of Europe. In former times this **situation** was unfavourable for trade. When the commerce of the world was carried on chiefly between the countries bordering the Mediterranean Sea, Britain was obviously quite outside the main streams of trade. It was at the edge of the world, and what little trade it had was hampered by such a position. But nowadays its situation with regard to the other great countries of the world is one of its most important assets. The commercial centre of gravity has shifted from the Mediterranean Sea to the Atlantic Ocean. The world's greatest trade route is between the English Channel and the United States, and the British Isles have plainly **the finest situation in Europe for Atlantic trading.**

The "world position," too, of the British Isles is remarkable. In classical times, as we have said, the British Isles were at the edge of the world. Now when all the land of the globe is known, it is seen that our country is very nearly at the **centre of the land hemisphere**[1], that is, the

---

[1] Authorities differ as to the exact position of the centre of the land hemisphere. Thus Gregory (*Structural Geography*) places it at London, Penck (*Morphologie der Erdoberfläche*) places it south-west of Paris; and De Lapparent (*Géographie Physique*) places it at Berlin.

half of the globe that contains more land than any other hemisphere that may be selected.

The British Isles are well placed, too, with regard to the chief countries of the Continent of Europe. They lie **close to the mainland**, only a few hours' travel dividing them from Germany, France, Holland, and Belgium, and are therefore able to share in the trade that is continually going on between all the important countries of Europe. It is true that Britain is not so well situated in this respect as Germany, for the latter touches most of the richest countries in Europe; but still Britain's position is better than most. Then Britain has the advantage over all the other countries of Europe in being **surrounded by the sea**. There is therefore no necessity for a huge standing army, and so more of the inhabitants are left free for peaceful, industrial pursuits which add to the wealth of the country. Notice, too, the point at which England approaches the Continent most closely. It is just where the **Celtic and Teutonic races of Europe meet**, and these two streams of influences have mingled intimately both in the blood and in the civilization of the British race, with the good results that follow all judicious "crossing." Remember, too, the great geographical truth that an island home tends to produce a vigorous civilization. Thus influences, both of blood and culture, in other words, heredity and environment, have helped to produce a self-reliant, enterprising race.

The geographical position of Britain involves other advantages. The **climate** is not so mild and genial as to put a premium on laziness, nor on the other hand is it so severe as to prevent out-door work in any season. As we shall see later on, the winters are exceptionally mild considering the latitude, and inland navigation is seldom impeded by ice, as it is on the Continent. Again, the British coasts abound in **good harbours**, and no part of the

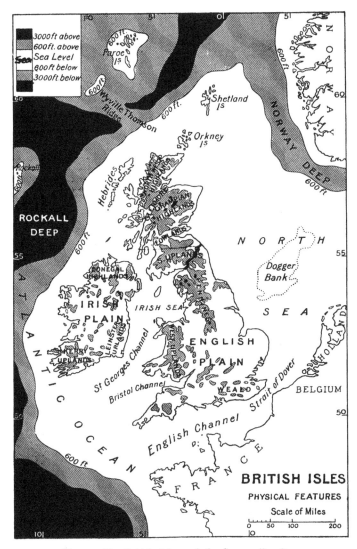

Fig. 1. The British Isles and the Surrounding Seas.

4 THE POSITION OF THE BRITISH ISLES

country is very far from a seaport. The country, too, is **rich in minerals,** and particularly in the most important of all minerals, namely, coal. Coal is the basis of the industrial wealth of to-day, and no other country in Europe equals Britain in the quantity or the quality of its coal output.

Turning now from the general aspect of the situation of Britain let us consider next some of the more important details regarding the relation of these islands to the neighbouring seas and lands. The seas around the British Isles are shallow. The submarine slopes of Europe towards

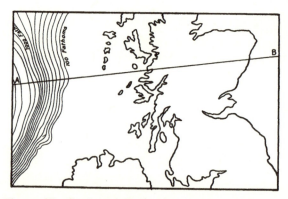

Fig. 2. The Continental Shelf. The contour-lines (drawn at intervals of 100 fathoms) show the sudden descent at the western edge of the Shelf.

the west are very gentle at first. This gentle slope is maintained for some distance to the west of Ireland before the sea-floor plunges more steeply to the depths of the Atlantic Ocean. The submarine area between the shores of the Continent and the steeper slope of the sea-floor is called the **Continental Shelf.** The British Isles therefore rise from a Continental Shelf. Quite the most striking way of realising this sudden steepening of gradient is by applying our knowledge of contour-lines to the interpretation of the slopes of the sea-floor. Look at Fig. 2 which shows the

THE POSITION OF THE BRITISH ISLES 5

bed of the ocean contoured at intervals of 100 fathoms. Notice just west of the 100 fathom line how the contour lines are crowded together showing that the steepness of the sea-floor there suddenly increases. It should be remembered that when we say the slope west of the 100 fathom line is steep, we mean steep compared with the slope of the Continental Shelf. In reality the steepest part of the slope is about 1 in 18, that is, roughly equivalent to a gradient at which an ordinary cyclist would dismount and walk his bicycle uphill, but which could be surmounted by a strong rider. The slope of the Continental Shelf could not be perceived by the unaided eye. In section the slope of the sea-floor west of the British Isles is seen in Fig. 3.

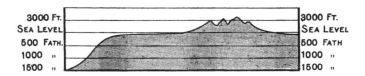

Fig. 3. Section across the Continental Shelf and Scotland. The section is drawn along the line *AB* in Fig. 2.

If we consider the Continental Shelf to be bordered by the 100 fathom line, we see that the seaward margin of the submarine platform keeps well to the west of the British Isles, being from 20 to 80 miles west of Ireland, and about 60 miles west of the Outer Hebrides. North of Scotland it swings round the outside of the Shetland Isles, and then one would expect it to keep north-east to the Norwegian coast. At this point, however, the Shelf is trenched by the remarkable submarine valley called the Norwegian Depression (see Fig. 1). This long and deep channel keeps close to the Norwegian coast, and swinging round to the north is continued into the heart of Scandinavia by the Christiania Fiord and the valley of Lake Miosen.

Most of the North Sea is exceedingly shallow. Many parts of the Dogger Bank, more than a hundred miles from land, are only ten to twenty fathoms deep. Indeed the depth of the North Sea is no greater in proportion than the thickness of a sheet of notepaper is compared with its area. The Irish Sea is also shallow. This sea, however, is crossed from north to south by a submarine valley that can be traced continuously through the North Channel, across the Irish Sea west of the Isle of Man, and southward along the middle of St George's Channel. A somewhat similar submarine valley is found farther north between the Outer and the Inner Hebrides.

The **shallowness of the seas** round the British Isles is important in two ways. In deep water the height of the tidal wave is insignificant. In mid-ocean high tide means only a rise of two or three feet. When the tide comes in over a continental shelf, however, the rise is very considerable. For example, in the Bristol Channel there is sometimes a difference of forty feet between high and low tide. It is therefore the shallowness of the seas round our coasts that gives us the high tides that are so useful in cleansing our shores, in filling our docks, in scouring our channels, in floating barges in and out our estuaries, and in bringing our great ships safely to port in deep water. Again, fish congregate and feed mainly in shallow water and over submarine banks. The banks of the North Sea are among the most valuable fishing grounds in the world, and many thousands of British fishermen obtain a livelihood from them.

The **tidal wave** that impinges on the British Isles comes from the south-west. Near Ireland it divides into three streams. One flows up the English Channel, another into the Irish Sea by St George's Channel, and the third keeps west of Ireland and flows round the north of the British Isles. The last named tidal wave moves fastest, and

THE POSITION OF THE BRITISH ISLES 7

entering the North Sea from the north it has reached the estuary of the Thames before meeting the tide through the English Channel. This is a fact of economic importance, for the meeting of the two tidal waves results in exceptionally high tides in the estuary of the Thames.

If the level of the sea round our shores were to sink 200 feet, England would be joined to the Continent. The English Channel and the North Sea south of the Dogger Bank would become dry land. The Irish Sea would become a great plain trenched north and south by a long inlet of the sea occupying the submarine valley referred to above. It is practically certain that the **British Isles once formed a part of the Continent.** The earliest men who inhabited Britain reached the country before the English Channel was formed. Many different lines of evidence point to this conclusion. The chalk cliffs of France have every appearance of having been once continuous with the chalk cliffs of England; the Fen district corresponds with the low-lying parts of Holland; northern Scotland is identical in structure with Norway. Again, the animals of England and the Continent are almost the same, although England has fewer kinds than the Continent, and Ireland still fewer; as if the animals of the Continent had not all had time to migrate to England and Ireland before the formation of the Irish Sea and the English Channel. For example, the famous naturalist, Alfred Russel Wallace, tells us that Germany has 90 species of mammals, Great Britain has 40, and Ireland has 22. Similarly Belgium has 22 species of reptiles and amphibia, Great Britain has 13, and Ireland has 4. The same story of a former extension of the land is told by dredgings from the North Sea. Off the east of Scotland there is a submarine bank from which shells are obtained of a kind that belonged to animals that live only at sea level. From the Dogger Bank have been dredged many bones of the mammoth, the rhinoceros, the

8 THE POSITION OF THE BRITISH ISLES

reindeer, and other land animals, showing that the North Sea was once dry land. We may conclude, therefore, with reasonable certainty, that the British Isles once formed part of the Continent, and that primitive man once hunted his prey where now the billows of the sea roll far from any land.

Not only were the English Channel and the North Sea dry land, but the **British Isles once extended much farther to the west**. It is very important to realise this clearly, for, as we shall see later, many of the most puzzling features of British geography are explained by the fact that this country once extended farther to the west. The waves of the Atlantic Ocean toss their crests over part of foundered Britain. At present it will suffice to mention one feature of the geography of Britain that is directly connected with the foundering of the western part of a continent more extensive in former times than now. Look at the west coast of Scotland. What a remarkably broken and indented outline is presented by this coast! How numerous, too, are the islands of the west coast! Then look at the east coast and note the contrast. It is sometimes said that the indented western coast has been produced by the vigorous breakers of the Atlantic, but this is an absurd explanation. Wave action could never produce these long, narrow, deep fiords. This broken coast line is probably a result of the fracturing of land that once existed far to the north-west, but has now sunk beneath the sea.

Let us sum up now the main points of this chapter. The destinies of Britain have been shaped in no small measure by its situation as an island group, just to the west of that part of the Continent where diverse streams of culture meet. Favourable position is further aided by good climate, many harbours, and rich natural resources. The islands rise from a continental shelf that west of Britain slopes more rapidly to oceanic depths. This fact, too, is shown to involve favourable results to the inhabitants of this country. Many reasons force us to the conclusion that Britain was once part of the Continent of Europe from which it has been sundered by a sinking of the land or a rising of the sea. The British Isles once extended much farther to the west, but the land was fractured, and much of it sank beneath the waves. The foundering of part of a former land probably took place before Britain was inhabited by human beings, but palaeolithic man lived in this country before the separation of the British Isles from the Continent.

CHAPTER II

THE BUILD OF THE BRITISH ISLES

FROM any good orographical map a great deal of information can be obtained without the use of any book whatever. It is advisable first of all to obtain familiarity with the scale of the map by making a few measurements. For example, find the distance between Duncansby Head and Land's End, between London and Carlisle, between Dover and Calais, between Fishguard and Rosslare, between Holyhead and Dublin, between Stranraer and Larne. Find the width of Scotland at its broadest and at its narrowest parts. By using transparent squared paper an estimate may be made of the area, say, of Ireland. It is somewhat less than 33,000 square miles. The United Kingdom is one of the smallest of the great powers of the world. It is a little more than half the size of Germany, and only one twenty-fifth part of the size of the United States.

Now note the meaning of the colour scale adopted for

the map, and then verify the following account of the configuration of the country. In Great Britain there is a strong **contrast between north-west and south-east.** Imagine a line drawn from Newcastle through Leeds, Sheffield, Derby, the Severn estuary, and Exeter. South-east of that line Great Britain is a land of plains. What hills there are, are low. North-west of the line is a mountainous country with intervening valleys. The contrast is fundamental. It is not confined merely to a matter of height above sea-level, but has a basis in rock structure, and finds an expression in the vegetation, the industries, and the very nature of the people.

In Great Britain there are **three great mountain masses** and a number of smaller ones. The first mountain mass forms roughly the northern half of Scotland, and is called the Highlands. The Scottish Highlands are divided into two parts by the long, straight, and narrow valley of Glen More. On one side are the North-West Highlands, on the south-east of Glen More are the Grampian Highlands. The Highlands are divided from the next hill mass by the Central Lowlands of Scotland. South of the Central Lowlands are the Southern Uplands which stretch completely across Scotland from north-east to south-west. This hilly region extends across the border into England, and is continuous with the Pennine Uplands which stretch to the southward for 150 miles. The Lake District Mountains must be considered as forming part of this mountain mass, for they are attached to the Pennines by a neck that is nowhere much less than 1000 feet in height. The third mountainous area comprises most of Wales. In the south-western peninsula of England there are several hilly areas of considerable extent, although not to be compared with those just mentioned.

The hilly regions described above are all situated north or west of the line from Newcastle to Exeter. South-east

THE BUILD OF THE BRITISH ISLES

of this line there are certainly hills, but they are much lower, much smaller in area, and of a different nature. The hills of south-eastern England nearly all consist of **long ridges** due to the outcrop of hard bands of rock. A limestone ridge extends from the head waters of the Thames to northern Yorkshire. A ridge of hard chalk can be traced from the Thames west of Reading to the Wash, thence to the Humber, and thence to Flamborough Head. There are other chalk ridges in southern England of a similar character. The map alone is sufficient to show that these ridges form quite a different type of hills from those of the north and west. The ridges are rarely a thousand feet in height and run in long lines for many miles. The hill regions of the north-west on the other hand are great massive blocks of rock that have been carved into a confused area of high mountains separated by deep valleys.

It is obvious that the **main watershed** of the country must be much nearer the west coast than the east. As a rule the eastern slopes are long and gradual while the western slopes are short and steep. For example, consider the highest peaks in Scotland, England, and Wales. Ben Nevis is only four miles from the western sea, Scawfell and Snowdon are but ten miles from the coast. This is certainly a most remarkable circumstance, particularly when we remember that erosion is more vigorous on the western slope than on the east. How can we explain it? The explanation lies in the fact that Great Britain is only the fragment of a country. The western coasts represent the broken edge, and it follows **that the land must be higher there than along the eastern margin**. If this is not clear, think of an island rising gently on all sides from the sea to the central summit. If this island is broken across the middle, plainly the fractured edge will be higher and steeper than the unbroken coast.

12 THE BUILD OF THE BRITISH ISLES

In Ireland the arrangement of high and low ground differs from that in Great Britain. The centre of the country is occupied by a great plain while the hilly regions are for the most part found near the coast. Long ridges, like those occurring in England and caused by the outcrop of hard bands of limestone or chalk, are not found. Formerly, in Ireland, the arrangement of high ground in the west and low ground in the east was also clear. But the eruption of great masses of igneous rock in the east has altered this arrangement.

The mountains of Britain differ profoundly from such systems as the Alps or the Himalayas. The latter have been produced by the wrinkling up of the solid crust of the earth by gigantic thrusts, acting very slowly and gradually. Mountains formed in this way run in long parallel ranges divided by long parallel valleys. Ranges of this kind are called folded mountains. They certainly existed at one time in Britain, but have long since been levelled by denudation.

The mountains of Britain do not show any true ranges. If we were to take our stand on a mountain top in the heart of the Scottish or Welsh Highlands, and look around, we should see no parallel mountain ranges, but a rough and tumbled sea of peaks, without any apparent order or arrangement. The mountains of such an area have been carved by the rain, the frost, and the rivers, out of a plateau, and if the material worn out of the valleys were replaced, such an area would again be a tableland. A mountainous area of such a nature is known as a **dissected plateau**. All the large mountain masses of the British Isles are of this type.

But the question arises, why should the hilly areas of north-west Britain not have been reduced by denudation to the state of plains such as are found in south-east Britain? The answer is very simple. They are made up of harder

THE BUILD OF THE BRITISH ISLES 13

and more resistant rocks than are found in the south-east, and rain, rivers, and frost have not yet had time to wear them down to the state of plains. The **hills** of the

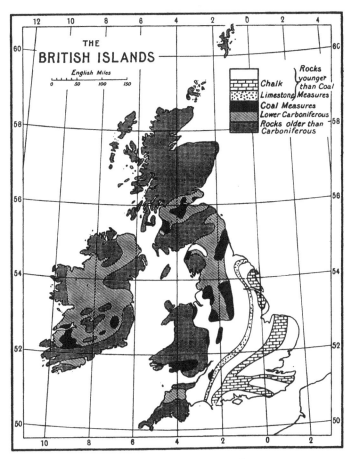

Fig. 4. Sketch-map of the British Isles showing the nature of the rocks.

British Isles are composed of **hard rocks**, the **plains** are composed of **soft rocks**.

The enquiry may be pushed a little further, and it

may be asked, why should some of the rocks be so much harder than others? The answer involves a slight knowledge of very simple geology. Most of the rocks of this country were once sediments on the sea bottom. The solid crust of the earth is largely made up of such sediments, and it follows that the oldest rocks will be overlaid by rocks that are successively younger and younger. For millions of years this process of rock accumulation has been going on, with intervals when the rocks were raised into dry land, and perhaps folded into mountain ranges. The oldest rocks have been most subjected to heat and pressure and other consolidating agents. They are therefore more compact, more crystalline, harder, tougher—in a word, they are more resistant to denuding agents.

The rocks that make up the crust of the earth are divided by geologists into about a dozen systems. One of these systems—the **Carboniferous System**—is especially important, for it contains the coal that is so necessary to the industries of Britain. The lower half of the Carboniferous System contains thick beds of limestone, found in many parts of the Central Plain of Ireland and also in the Pennines. The upper half of the system is called the **Coal Measures**, for it contains many seams of coal as well as valuable beds of ironstone, fire-clay, sandstone, and other rocks. The black areas in Fig. 4 represent the Coal Measures.

We can now see how close a connection there is between the nature of the rocks and the relief of the British Isles. Look carefully at Fig. 4. In this map we have shown the Carboniferous System because of its special importance in containing the Coal Measures, and also because we have taken it as the division between older rocks and newer rocks. It is at once apparent from this map that practically all the mountain masses of the British Isles are composed of old rocks, while the plains consist

THE BUILD OF THE BRITISH ISLES 15

of young and less resistant rocks. The Scottish Highlands, the Southern Uplands, the Lake District, Wales, the hills of Cornwall and Devon, the Donegal mountains, the Wicklow mountains, the Kerry mountains—all the important hill masses of Britain consist of old rocks. The Carboniferous System itself sometimes occurs as hills, sometimes as plains. In the Pennines and part of the south-west of Ireland these rocks form hill masses. In the Central Plain of Ireland and the Lowlands of Scotland they form low ground. Among the younger rocks, however, there are two bands of special importance, one of **hard chalk**, the other of **hard limestone**. These bands are more resistant to rain and rivers than the softer rocks in which they occur, and therefore where these bands outcrop at the surface we find lines of hills. The Cotteswold Hills, the Cleveland Hills, the Chiltern Hills, the North Downs and the South Downs are formed by the outcrop of hard bands, either of limestone or of chalk.

CHAPTER III

WEATHER AND CLIMATE. WINDS

THE weather of Britain depends largely on the distribution of atmospheric pressure over these islands. To put the matter in its simplest form, when the barometer is high we expect good weather, and when the barometer is low we expect wet and stormy weather. These two types of weather correspond respectively to a condition of high atmospheric pressure or **anticyclone**, and a condition of low atmospheric pressure or **cyclone**. Look at Fig. 5, which shows the isobars and directions of the wind on a day in

16 WEATHER AND CLIMATE. WINDS

winter. It is plain that over the British Isles a cyclone is passing, the centre of which lies to the north of Scotland. The figure represents a type of weather conditions very frequently experienced in this country. The wind directions are shown by arrows. The arrows show that the wind does

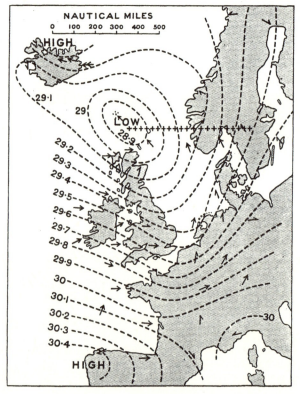

Fig. 5. Pressure Chart illustrating a Cyclone. November 22, 1908.

not blow quite parallel to the isobars, nor directly towards the centre of the cyclone, but combines a motion round the cyclonic centre in a direction opposite to the hands of a clock with an inward motion towards the centre. If the

WEATHER AND CLIMATE. WINDS 17

air is swirling round with an inward tendency, where does it go to, for the spiral motion towards the centre lasts for days? It is plain there must be a continuous upward motion of the air into higher parts of the atmosphere. As the air rises it is cooled, and thus there is a tendency for

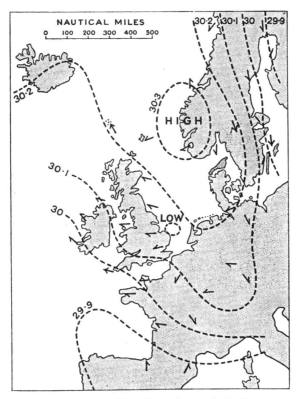

Fig. 6. Pressure Chart illustrating an Anticyclone. November 6, 1908.

the moisture to be condensed. **Rain, therefore, generally accompanies a cyclone.** The likelihood of rain, too, is increased by the fact that the centre of the cyclone generally lies to the north of Britain, and so south and west winds are prevalent. These winds come over a warm ocean and therefore tend to bring rain. Notice again that the barometric gradient is fairly steep. Contrast this figure with the next in that respect. This means that the **winds are strong**. Cyclones generally move from west to east. The right-hand end of the line composed of small crosses shows where the centre of the cyclone was on November 23rd, twenty-four hours later.

Consider next Fig. 6, which shows an anticyclone over the British Isles. In this case the distance between the isobars is much greater than in Fig. 5, showing that the barometric gradient is very slight. The **winds, therefore, are very light**, and this is characteristic of anticyclones. There is a gentle circulation outwards and also in the same direction as the hands of a clock. Since the air is moving outwards there must be a descending current. The descending air comes from colder parts of the atmosphere to warmer, and therefore its humidity is diminished. Thus **fine, clear, sunny weather** is associated with anticyclones. In summer the weather is sunny and hot, in winter it is clear and frosty. In low-lying ground and in towns the absence of wind involves **mists** and **fogs**. Anticyclones do not move in definite tracks like cyclones but gradually disappear.

Generally speaking we may say that the winds of the British Isles are controlled by **three** fairly permanent **pressure centres**. There is a low-pressure area south of Iceland, an Atlantic high-pressure area about the Azores, and a continental area in eastern Europe and western Asia that is high in winter time and low in summer time. The frequent cyclones that invade our coasts seem to be smaller eddies thrown off by the permanent Icelandic low-pressure area. In winter as a rule the Icelandic low-pressure centre and the continental high-pressure centre predominate. They are then working in harmony, for the British Isles

WEATHER AND CLIMATE. WINDS 19

are between the Icelandic cyclone which is drawing the wind counter-clockwise from south-west to north-east and the continental anticyclone which is sending the wind clockwise from south-west to north-east. The tendency of both centres, therefore, is to draw the air in a great swirl between them from south-west to north-east. Thus we find that **in winter south-west winds predominate** over the British Isles. Occasionally the continental anticyclone gets the upper hand and spreads over these islands. Then for a few days in winter we experience clear skies, keen frosts, and very light winds. All too soon the Icelandic cyclone centre reasserts its sway, and we are back again to storms of sleet or rain with a higher temperature.

In summer the Atlantic high-pressure centre has more influence. This area with its accompanying fine weather is now at its most northerly limit, for to some extent it moves north and south with the sun. In addition, the Icelandic and the continental centres are no longer in harmony, for both are now cyclonic and Britain lies between them. The anticyclone occasionally spreads over these islands, reaching the south of England frequently, but not so often extending to Scotland. The Atlantic anticyclone tends to draw the winds more to the west, sometimes even to the north-west. This shift of the winds from south-west in winter to **west in summer** is illustrated by the wind roses on p. 20.

We have stated that the winds shift from south-west to west according to the season, and we have offered an explanation of this shift. Let us see now if actual observations confirm the statements we have made. It will not do to select one year only, but an average over several years. We shall show the average wind directions for sixteen years, from 1893 to 1908 inclusive. Instead of giving numerical tables we have expressed the results as diagrams from which the prevailing winds may be seen at a glance.

20 WEATHER AND CLIMATE. WINDS

Along each of the eight principal points of the compass we mark a distance proportional to the percentage of days on which the wind blew from that direction, and so get a star, the longest points of which show the winds that blow

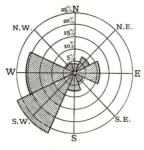

Fig. 7. Wind Rose showing the frequency of the winds from the eight principal points of the compass during January.

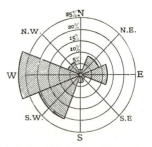

Fig. 8. Wind Rose showing the frequency of the winds from the eight principal points of the compass during July.

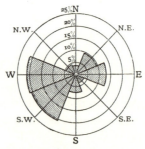

Fig. 9. Wind Rose showing the frequency of the winds from the eight principal points of the compass throughout the year.

most frequently. Fig. 7 shows that the winds of winter are chiefly from the south-west, and Fig. 8 shows that the winds of summer are chiefly from the west. In **late spring and early summer easterly winds** are fairly common.

WEATHER AND CLIMATE. WINDS

This results from the fact that in spring cyclones frequently travel along the English Channel or the Bay of Biscay. A little consideration will show that this must result in cold east winds from the Russian plains being drawn over the British Isles. Fig. 9 shows the prevailing winds for the whole year. West and south-west winds are clearly the most common. In many parts of the country the trees are silent but reliable witnesses to the same fact. They grow with their branches pointing east or north-east, away from

Fig. 10. Tree showing effect of prevalent south-west winds. The branches point north-east. The photograph was taken in a calm.

the wind. The branches of the trees shown in the photograph above (taken on a still day) point almost exactly north-east.

It is still quite generally believed that storms are more frequent and violent at the time of the equinoxes than at other times. The phrase "equinoctial gales" is heard so frequently that people come in time to believe that September and March are especially stormy months. There is no foundation for the belief. **The equinoctial gales are** mythical. In order that there can be no doubt about the matter let us take a period of forty years, 1868—1907, and note the number of gales in each month, recording all storms that attained over forty miles an hour. These are the numbers for each month during that time:

Jan.	Feb.	Mar.	Apl.	May	June	July	Aug.	Sept.	Oct.	Nov.	Dec.
50	42	36	11	5	2	2	5	10	15	27	39

These figures show clearly that storms are most frequent in winter and least frequent in summer. The maximum number occurs in January, and the number steadily decreases till June and July, then rises steadily to January.

Let us sum up now the chief points of a rather dry chapter, which, however, if thoroughly grasped, will explain some of the most outstanding features of the weather of this country. The two principal types of weather are cyclonic and anticyclonic, the former being unsettled, stormy, and wet, the latter settled, dry, and sunny, hot in summer, cold in winter. Apart from frequent minor cyclonic disturbances the weather of Britain is controlled mainly by three pressure centres, a low-pressure area south of Iceland, a high-pressure area near the Azores, and a continental area low in summer, high in winter. These pressure centres cause the winds of winter to be mainly from the south-west, and of summer to be mainly from the west. The so-called equinoctial gales are a meteorological myth, without any basis of accurate observation. Storms are most frequent in winter and least common in summer.

CHAPTER IV

WEATHER AND CLIMATE. RAINFALL AND TEMPERATURE

THE distribution of the rainfall over the British Isles is shown in Fig. 11. A glance at the map makes plain two fundamental facts concerning British rainfall, first that **the east of Britain is drier than the west,** and second that **the greater the altitude the heavier the rainfall.** The first fact is explained by what we learned in the previous chapter, namely that the prevailing winds of Britain are west and south-west. They blow over a warm ocean and the moisture they carry is condensed largely on our western coasts. The correspondence between higher altitude and heavier rainfall is very striking. The rainfall map in many respects resembles an orographical map. For example, the highest parts of Great Britain are the Scottish Highlands, the Lake District, and Wales, and these are the rainiest parts of the country. Again in Ireland, notice how the Mourne Mountains and the Antrim Plateau are represented on the rainfall map almost as accurately as on the orographical map. Notice also how clearly the North Downs and South Downs stand out by reason of their higher rainfall, and how the Cleveland Hills and the Yorkshire Wolds can likewise be identified. Try to find from the map other correspondences between altitude and rainfall. Let us illustrate the same point by one numerical example, probably the most striking in Britain. In Fort William at the foot of Ben Nevis the average annual rainfall is 73 inches, while at the top of the mountain the average rainfall is about 160 inches per annum.

The rainfall is not uniform throughout the year. **Most rain falls in winter time.** In London, November is the rainiest month, in other places (the west of Scotland, for example), January has most rain, but everywhere in Britain winter is the rainy season. Spring is the driest season of the year. March, April, and May are as a rule the months with least rainfall. As most holiday makers know, there is generally a distinct rise in the rainfall in late summer, July and August often being rather rainy months. The difference in rainfall between the seasons, however, is not great enough to justify an adjective such as "seasonal" as applied to the rainfall, for rain falls in every month in the year.

The maps on p. 27 show the mean temperature of the British Isles in January and in July. The lines on the maps join places having the same mean temperature for these months, and are therefore called isotherms. In the first place it should be noted that the actual mean temperature of hilly regions is not shown on these maps, for isothermal maps show what the mean temperature would be if all the country were at sea-level. But it is quite easy to find what the actual mean temperature of any place is if we know its height above sea-level. There is a fairly constant fall in temperature as one goes above sea-level, amounting roughly to 1° F. for every 300 feet of ascent. For example, the January isotherm of 39° F. passes very close to Ben Nevis. But the summit of that mountain is 4400 feet above sea-level and therefore the actual mean temperature of the summit is roughly 24° F.

Now from an examination of the maps determine the answers to the following questions: What part of Britain is warmest in summer? What part of Britain is coldest in winter? What part of Britain has the most extreme climate? If we pass from London to Land's End in summer, does the temperature rise or fall? Does the same

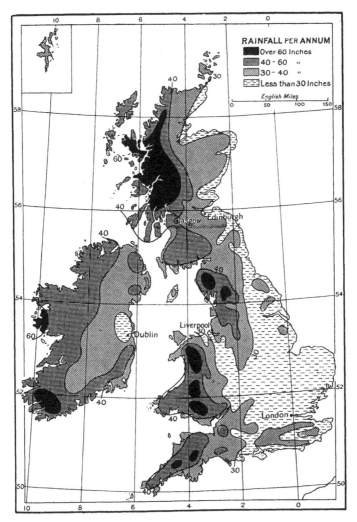

Fig. 11. Rainfall Map of the British Isles. (After Mill.)

thing hold good in winter? Now let us compare pairs of places in approximately the same latitude, one place of each pair being on the east coast, the other on the west coast. Write down in tabular form, to the nearest degree, the January temperature, the July temperature, and the mean annual range of temperature for the following pairs of places: Land's End and London, Lewis and Aberdeen, Dublin and Galway. These questions and exercises make plain certain important facts regarding the climate of the British Isles. The **most extreme** climate is found in the **south-east of England.** This part of the country is coldest in winter and warmest in summer. Remember also that we have already seen that it is the driest part of Britain. Again in all parts of the British Isles,—England, Scotland, and Ireland,—the west coast is more equable than the east. This is a climatic principle of wide application. For example, if we were dealing with all Europe instead of the British Isles, we should find a mean annual range in the extreme west of Ireland of about 17° F. and in the extreme east of Russia of no less than 67° F. Notice how the isotherms in crossing the Irish Sea bend northwards in January and southwards in July. This shows that the sea is warmer in winter than the land, and in summer is colder.

As regards temperature the climatic conditions of the British Isles are abnormal. **Winter** in the British Isles is **milder** than in any other part of the world in the same latitude. How favourable is our situation in this respect is strikingly shown by comparing Britain with eastern America. Aberdeen and Nain (Labrador) are in the same latitude. The mean temperature of the coldest month at Aberdeen is 37° F. or five degrees above freezing point. The mean temperature of the coldest month at Nain is $-4°$ F., that is, thirty-six degrees below freezing point. The British Isles lie in the centre of what has been called

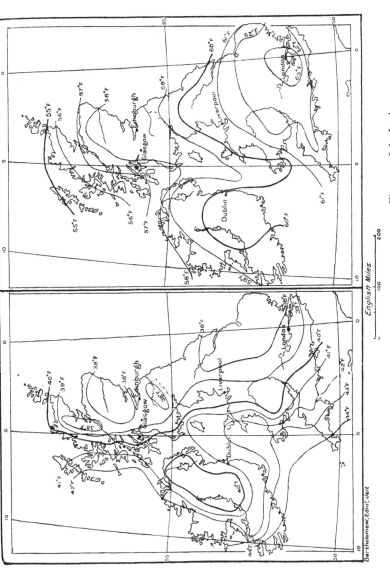

Fig. 12. January Isotherms.

Fig. 13. July Isotherms.

a "gulf of warmth." In winter a great V-shaped area, with comparatively warm weather, extends far up the western coasts of Europe. The undotted part of Fig. 14 shows how this abnormally mild area extends past the British Isles and far to the north-east.

If we look again at the figure showing isotherms for January, we shall see that the existence of some unusual climatic factor is shown by the **abnormal direction of the isotherms**. As a rule isotherms run roughly from east to west because places in the same latitude, other things being equal, have approximately the same mean temperature. But the January isotherms for the British Isles actually run north and south. It is warmer at Cape Wrath in winter than it is in London.

Until recently it was believed that our unusually mild winter climate was due to the beneficent influence of the Gulf Stream, but in recent years this explanation has been entirely abandoned. It is a myth as fanciful as the supposed equinoctial gales. The Gulf Stream as a well-marked current becomes a negligible climatic factor a little to the east of the Newfoundland banks. Our true benefactor is the wind. We have already seen how persistent are the south-west winds of winter. They blow from the warm southern regions of the Atlantic Ocean, raising the temperature of Britain and depositing moisture, which owing to the liberation of the latent heat means a still further rise of temperature. In addition they blow the warm surface waters of the ocean from more southerly latitudes, and cause them to flow round and past our islands. There is no strongly marked current but a general surface drift of the warm upper layers of the water. This motion of the sea is usually known as the "**Atlantic Drift**," or sometimes, though less accurately, as the "Gulf Stream Drift."

From what has been already said in this and the preceding chapter it will be plain that the amount of sunshine received by different places in the British Isles will vary considerably. The southern coasts of England are most fortunate in this respect. On an average they have about 1600 hours of sunshine in the year, while

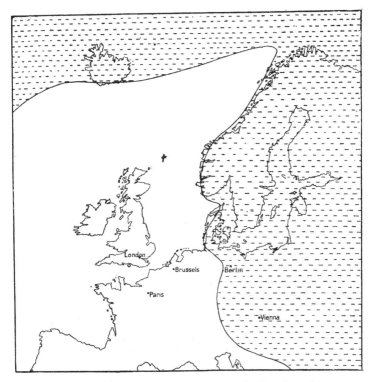

Fig. 14. Gulf of Warmth over western Europe. The dotted area has a mean temperature during January of less than 32° F. (After Mackinder.)

parts of the Scottish Highlands have less than half that amount.

Summing up this chapter, then, we find that the rainfall of the British Isles increases as we pass from east to west, and to even a greater degree as we rise above sea-level.

So pronounced is the latter fact that a rainfall map presents many resemblances to an orographical map. Rain falls in Britain in every month in the year, although in most parts there is a winter maximum. The west coasts of this country are milder in winter and cooler in summer than the east. The south-east of England has the most extreme climate of any part of the British Isles. These islands are situated in a "gulf of warmth" caused by the warm waters of the Atlantic Drift. No other country in the same latitude has as favourable a winter climate as the British Isles. The south coast of England is the sunniest part of Britain.

CHAPTER V

AGRICULTURE AND FISHERIES

THE character of the vegetation of a country depends mainly on two things, the climate and the nature of the soil. We have already seen that the mountainous parts of Britain are much colder, have a much higher rainfall, and are much more exposed to wind than the lowlands. In addition, the soil is poor, for the rocks are hard and crystalline, and do not weather into deep, rich soils. For these reasons neither crops nor trees can be grown much above sea-level. The **hilly parts** of the United Kingdom then are mainly **uncultivated**, and a map distinguishing between the cultivated and the uncultivated parts of this country would present a striking resemblance to an orographical map. The basic factor is of course the nature of the rocks, for that determines the altitude and the nature of the soil,

AGRICULTURE AND FISHERIES

and the altitude controls the climate. Besides the mountainous parts there are areas of boggy and sandy ground which are not cultivated.

The cultivated portion of this country includes those parts that are under crops and also the area of grass land used for pasturing animals. Roughly sixty per cent. of the United Kingdom is cultivated, the rest being scanty mountain pasture, heathy land, sand, bog, or bare rock.

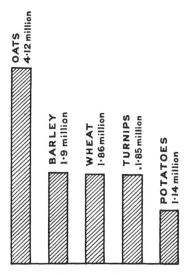

Fig. 15. Diagram showing comparative areas under the chief crops in 1910.

In the different countries, however, the proportions of cultivated land vary considerably. About three-quarters of England and Ireland are cultivated, about sixty per cent. of Wales, and only a quarter of Scotland. Of the cultivated portion of the United Kingdom the greatest part consists of pasture land, which forms about fifty-eight per cent. of the cultivated portion, that is, a little more than a third of the total area.

The principal grain crops of the British Isles are oats, barley, and wheat. In 1910 the area under these crops was as follows:

Oats	4,116,000 acres
Barley	1,899,000 ,,
Wheat	1,858,000 ,,

Not nearly enough **wheat** is grown in this country to satisfy the needs of British consumers, and every year large quantities of wheat and flour are imported from abroad. Taking an average over the three years 1907—1909 we find that Britain produced 21 per cent. of her needs during that period, and imported not far short of four times that amount. The following table shows the chief countries from which our imports of wheat are obtained.

Imports of Wheat and Flour 1910

Russia	28·9 million cwt.
Canada	20·3 ,, ,,
United States	18·1 ,, ,,
India	17·9 ,, ,,
Argentina	15·3 ,, ,,
Australia	13·7 ,, ,,

From all our British possessions we received 52·6 million cwt. and from all foreign countries we received 66·5 million cwt. The fact that we can obtain wheat very cheaply now from abroad has led to a gradual diminution of the area under wheat in this country. Thirty years ago the wheat acreage of the United Kingdom was nearly twice what it is now. This is shown in Fig. 16. In order to ripen properly wheat needs a warm, sunny, dry summer. In a former chapter we saw that the south-east of England best fulfils this requirement, and so we are not surprised to find that the district between the Wash and the Thames is the greatest wheat-growing region in Britain.

Oats can stand cold and wet weather much better than wheat, and are therefore more commonly grown in most

AGRICULTURE AND FISHERIES

parts of this country. In parts of Scotland and Ireland oats are the only grain that is cultivated. The **barley** that is grown in the United Kingdom is not very much used for food. Most of it is sold to brewers and distillers for the making of beer and whisky. Of the other crops **potatoes**

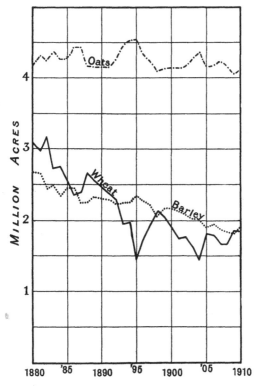

Fig. 16. Graphs showing areas under the chief grain crops from 1880 to 1910.

and **turnips** are the most important. Ireland grows more potatoes than England and Scotland put together.

Of the less important crops, **flax** is confined almost entirely to the north-east of Ireland. The local supplies, however, are not nearly sufficient for the demands of the great linen industry of North Ireland, and therefore large quantities of flax are imported, chiefly from Russia. **Hops** are grown mainly in the south-east of England. The sunny counties of Kent and Devon are noted for their **fruit**. The basin of the river Wye near the border of Wales is also famed for its orchards. It lies in the lee of the Welsh mountains, in their "rain-shadow," and the climate is consequently sunny and dry. In Scotland the middle part of the Clyde valley is one continuous fruit garden.

We have already mentioned that in recent years the area devoted to wheat has steadily diminished. The same is true for other crops, the result being that more and more arable land is being converted into permanent pasture, for many farmers are finding it more profitable to give increased attention to **stock-rearing** and **dairy-farming**. For example in 1880 there were in England 13 million acres of arable land and 11 million acres of permanent grass. In 1910 there were 11 million acres of arable land and nearly 14 million acres of permanent grass. The richest pastures are found in low-lying regions, particularly in those districts where the rainfall is high. From what we know of British climate, therefore, we shall expect to find the chief cattle-rearing districts in the rainy lowlands of the west. On referring to government returns we find that cattle are most important in Ireland, and in the counties of Cheshire, Somerset, Pembroke, Ayr, Renfrew, and Wigtown. Thus the location of the principal cattle-rearing and wheat-growing districts of this country affords us two excellent examples of the **control of occupations by climatic conditions**.

The chief **sheep-rearing** districts, too, are controlled by natural causes. Sheep do not require the same rich pasture land as cattle. They are much hardier and can thrive on

AGRICULTURE AND FISHERIES

scantier fare; although the hilly parts of the British Isles are too cold and wet for ordinary crops to ripen, there is often enough grass on the hill sides to feed sheep. So we find most sheep in hilly districts like the Downs, the Cotteswolds, the Chilterns, Wales, and the Cheviot Hills. The contrasted types of pasture land needed for sheep and for cattle are well illustrated in the cases of Scotland and Ireland. The grazing grounds of the former country are largely scantily-clad mountain slopes, the pastures of the latter are rich, low-lying meadows. Ireland possesses more cattle than sheep, while in Scotland sheep are seven times more numerous than cattle.

The **fisheries** of the British Isles are more important than those of any other country in Europe. The inhabitants of islands naturally tend to become good sailors and good fishermen. Then Britain is most conveniently situated to one of the richest fishing grounds in the world, the "banks" of the North Sea. Hundreds of thousands of tons of fish are caught every year by British boats, and the value of the catch amounts to millions of pounds. The herring catch is by far the most valuable, and haddock and cod rank next in importance. The following table shows the chief kinds of fish in order of value in 1910.

Value of Fish landed on coasts of U.K. in 1910

	Million pounds
Herrings	3·2
Haddock	1·8
Cod	1·6
Hake	0·6
Soles	0·4
Mackerel	0·3
Turbot	0·3

Total value of all kinds of fish 11·8 million pounds

Certain kinds of fish, such as the sole and the flounder, live near the bottom of the sea and are caught by trawlers which drag their nets along the bottom. Nowadays most trawlers are steam vessels. The great centres of the steam-trawling fleets are Hull and Grimsby which land more fish than all the other ports of England put together. Some kinds of fish, such as herring and mackerel, keep close to the surface as a rule, and these are caught by boats called drifters, which set their nets just below the surface of the water. Yarmouth is the most important centre of the fleets of drifters. Although the Scottish fishing grounds are not so valuable as the English, they are nevertheless of great importance and give employment to many thousand men. On the other hand the Irish fisheries are comparatively unimportant. In 1910 the values of fish landed in England, Scotland, and Ireland respectively were £8·2 million, £3·2 million, and £0·4 million.

CHAPTER VI

MINERALS

Coal is by far the most valuable mineral produced in this country, not only because the annual output is over six times more valuable than that of any other mineral, but because the industrial prosperity of this country is based mainly on its coal supplies. We are more purely a manufacturing nation than any other country in the world, and as such we depend entirely on coal as a source of power. There is no substitute available to take its place, and there is little likelihood of any being found in the near future. It is not too much to say that from the industrial point of view our coal supplies are our greatest national asset. The modern demand for coal in this country began about the end of

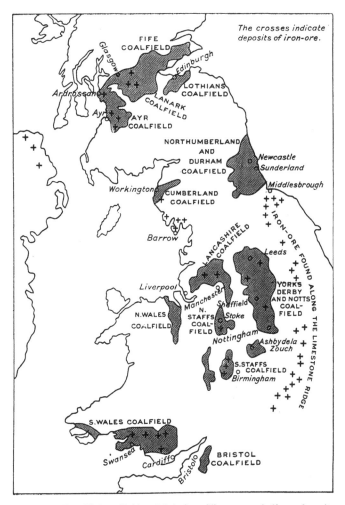

Fig. 17. The chief coalfields of Britain. The crosses indicate deposits of iron-ore.

the eighteenth century, when steam power was applied to all kinds of machinery. The textile trades were growing enormously, the iron-smelters had abandoned wood and were making greater and greater demands for coal, then came the application of steam to locomotion, both by land and sea, and later the use of coal gas for lighting, heating, and power. During all this time a great export trade in coal was being developed, and this has now reached immense proportions. At the beginning of the nineteenth century about ten million tons of coal were raised in this country; in 1910 the output had grown to the large total of 264 million tons.

The **principal coalfields** of the British Isles are shown in Fig. 17. Their names are as follows :—

England.
1. Northumberland and Durham.
2. Cumberland.
3. Yorks, Derby, and Notts.
4. Lancashire.
5. North Staffordshire.
6. South Staffordshire.
7. South Wales.

Scotland.
8. Fifeshire.
9. Midland (Lanark and Lothians).
10. Ayrshire.

Three of these coalfields have a much greater output than the others. These are the Yorks, Derby, and Notts, the Northumberland and Durham, and the South Wales. In addition there are a number of coalfields of less importance. In England there is one in North Wales, another near Bristol, and another at Ashby-de-la-Zouch. In Ireland there are several fields but none of great importance. There is one in south-west Ireland, another near Kilkenny, and another in Antrim. The ten large coalfields named above should be thoroughly known, for most of the

MINERALS

large towns of this country are situated on or near them, and they have attracted to their neighbourhood most of the manufactures of Britain.

The United Kingdom was the greatest coal-producer in the world until 1899, when our country was beaten by the United States which are now easily first in this respect,

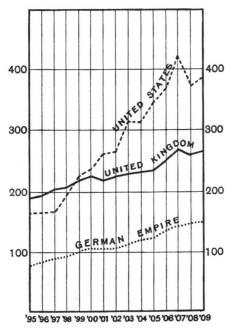

Fig. 18. Output (in millions of tons) of the three principal coal-producing countries of the world for the fifteen years 1895 to 1909.

Fig. 18 shows the output from 1895 to 1909 of the three leading coal-producers of the world. The surprising fact about Britain as a coal-producer is *not* that she has been surpassed by a huge country with enormous natural resources like the United States, but that she retains such a commanding position in the industry. This must be attributed to several special **advantages** that Britain enjoys **for coal production**. In the first place, compared with the size of the country the area of the coalfields is extraordinarily large, and again the quality of the coal is very good, much higher than that of most of the Continental fields. Further, facilities for handling the coal are exceptionally good. No field is inaccessible, and many of them actually touch the sea, such as the South Wales, the Northumberland and Durham, the Cumberland, and all the Scotch fields. Transport rates are therefore exceptionally light.

The amount of **coal exported** from this country has always been large. In 1910, 64 million tons were exported, nearly a quarter of all the coal produced in the country. France, Italy, and Germany are our largest buyers. In addition to this amount large quantities are sent to our own coaling stations in distant lands to be used by our vessels engaged in foreign trade. In 1910 this "bunker coal" amounted to twenty million tons. Fig. 19 shows in graphical form the amount exported during the fifteen years 1896—1910, bunker coal being ignored. The temporary drop in 1901 was due to the imposition of a tax on exported coal which was repealed in 1906.

One vital question for this country is the **duration of our coal supplies**. They cannot last for ever, and it therefore becomes of the highest importance to know approximately when exhaustion may take place. It seems certain that when our own resources of coal fail we shall sink to the level of a second-rate power, as far as industrial eminence is concerned. A Coal Commission made exhaustive enquiries into the subject and issued a report in 1905. By far the largest supplies of untapped coal were found in the South Wales and the Yorks, Derby, and Notts coalfields. We shall not go into detailed figures, but the final finding

MINERALS

of the Commission may be given. They stated that at the rate of production of 230 million tons per annum our coal would last for over 600 years. It should be noticed, however, that we have already exceeded that rate, and further that *cheap* coal will be exhausted long before that time.

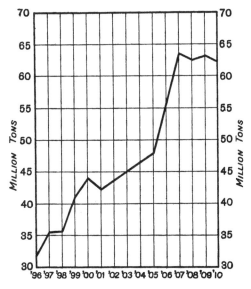

Fig. 19. Graph showing coal exports for the fifteen years 1896 to 1910.

After coal our most important mineral is iron. Although the iron ore of this country is gradually being worked out, yet millions of tons are still produced every year. In this respect we rank third among the countries of the world, for the United States and Germany produce more ore than we do. For the five years from 1905—1909 the annual production of the United Kingdom was about 15 million tons. In most of the coalfields deposits of iron-stone are found, generally in the form of carbonate of iron.

Valuable deposits of iron ore are found along the resistant limestone ridge that stretches from Gloucester to the north of Yorkshire, where it forms the famous Cleveland iron district, by far the most productive in Britain. Where it crosses the counties of Northampton and Lincoln the limestone ridge is also rich in iron ore. In South Cumberland and North Lancashire occur the rich haematite deposits of the Furness district. In Scotland, Ayrshire takes the first place as an iron producer, followed by Lanarkshire. In Ireland considerable quantities of ore are mined in Antrim.

In spite of the large quantities of iron ore raised in Britain our supplies are inadequate to meet the enormous demands of the great iron and steel industries of this country. We must therefore obtain supplies from abroad. The principal **places from which we import iron** are Spain, Elba, Sweden, and the north of Africa. The necessity of importing iron ores is a disadvantage, but not a serious one, for there is little land transport, and the cost of bringing the ore by sea is not heavy. In 1910 our imports of iron ore amounted to seven million tons.

In this country **tin** still ranks next to iron so far as the value of the annual output is concerned. For many centuries the British Isles produced most of the world's tin, the metal having been worked by the Phoenicians about three thousand years ago. But owing to the gradual exhaustion of the English mines and the discovery of rich fields abroad, chiefly in the Malay Peninsula, we now import much more tin than we produce at home. Cornwall and Devon are the only counties in which tin is worked. **Lead** is widely distributed, and is mined in Wales, Cumberland, the Pennine Uplands, and the Lowther Hills of Scotland. The amount of metal produced from home ores is not nearly enough to satisfy our needs and so much has to be imported. In fact less than one-tenth of all the lead

MINERALS

we use is produced in Britain. **Copper** and **zinc** are also widely found in the British Isles, but the amounts produced are almost negligible compared with what we import. Taking the case of copper as an example, we find that of the metal used annually in this country, the foreign-produced copper actually exceeds the British metal four hundred times.

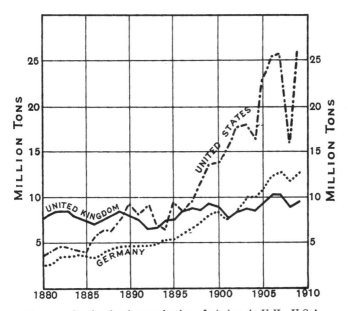

Fig. 20. Graphs showing production of pig-iron in U.K., U.S.A., and Germany from 1880 to 1910.

Great quantities of **salt** are mined in this country. Salt is largely used for flavouring and preserving food, but in addition it forms the chief raw material of many chemical industries. There is a rock series overlying the Coal Measures which is known as the New Red Sandstone, and practically all the salt mined in this country is obtained from these rocks. The commonest way of securing the mineral is to put down a bore hole and flood the salt-bearing strata with water. The water dissolves the salt, and the brine is pumped to the surface by steam-pumps. The final product is obtained by evaporation and refining. New Red rocks are found in Cheshire, Worcestershire, and the Cleveland district, and these places are the chief centres of the British salt industry.

Slates are found in Wales, the Lake District, and the Highlands of Scotland. The **materials used for building** depend to a large extent on the nature of the local rocks. London is fairly near good limestone quarries, so that many of the important buildings are of limestone. The towns in the Midlands find it more convenient to construct brick buildings, because clay is at hand in abundance while building stone is not. Glasgow is surrounded by fine sandstone quarries, and therefore it is built almost entirely of sandstone, while Aberdeen contains many granite buildings because it is situated in a granite district.

CHAPTER VII

THE BUILD OF SCOTLAND

WE shall now proceed to describe each country of the United Kingdom in some detail, and we shall begin with Scotland, because it presents a number of problems in physical geography which it is advisable to understand at the outset. An orographical map makes it plain that **Scotland consists of three parts**, a mountainous area in the north, a band of lower ground in the centre, and another hilly area in the south. These three parts are the Highlands, the Central Lowlands, and the Southern

THE BUILD OF SCOTLAND 45

Uplands, and they differ fundamentally from each other, not only in relief, but in vegetation, in the character of their scenery, in the nature of the underlying rocks, in their industries, and in their inhabitants.

The three divisions of Scotland do not merge gradually into one another. They are sharply separated by two lines running from north-east to south-west. **From Stonehaven** in Kincardine **to Helensburgh** on the Firth of Clyde there runs the Highland Line, the boundary between the Highlands and the Lowlands. **From Dunbar** in Haddington **to Girvan** in South Ayrshire runs the line dividing the Central Lowlands from the Southern Uplands. These two lines are of prime importance. They mark the course of two great cracks or "faults," between which the rocks of the Lowlands have gradually sunk, leaving high hill masses to the north-west and the south-east. Even to this day motion has not entirely ceased along these lines. From the districts near the Highland Boundary Fault reports of earthquakes are not uncommon, and these are caused by a slip or a readjustment of the rock masses along the fault. One other great fault in Scotland deserves attention. In a former chapter we stated that the Highlands were divided into two parts by the long, narrow, straight valley of Glen More. This valley is also caused by a fault, which like the others runs from north-east to south-west.

North of the Highland line the rocks are of a peculiar kind. They are called schists and gneisses, and are hard, crystalline, and resistant to the weather. The solid rocks have been bent, twisted, and folded in an amazing manner by stupendous earth forces. In places great slices of rock many square miles in area have been torn off and thrust bodily for miles over other rocks. Even in small specimens these rocks can be easily recognised by their crumpled appearance. Fig. 21 shows a typical specimen of a Highland

46   THE BUILD OF SCOTLAND

schist. The rocks of the Highlands are very old. In the Outer Hebrides and parts of the North-West Highlands we find the oldest rocks in Britain, the Hebridean Gneiss.

The rocks of the Central Lowlands are of a familiar type—sandstones, shales, and limestone. Most of them belong to the Carboniferous System, the upper part of which contains the valuable seams of coal that have made

Fig. 21.   A typical Highland Schist.

the Scottish Lowlands one of the busiest industrial parts of Britain. The Lowlands are not altogether flat. In various places hill masses rise to a height of 2000 feet above sea-level. These hills have the structure of plateaus. They consist of great tabular blocks of volcanic rock that have resisted denudation more than the surrounding rocks, and therefore stand out as hills.

THE BUILD OF SCOTLAND

The rocks of the Southern Uplands are intermediate in character between the rocks of the Highlands and those of the Lowlands. They are hard, but not so hard as the Highland rocks; they are folded, but not so much as the Highland rocks; they are old, but not so old as the Highland rocks. The Southern Uplands therefore resemble the Highlands but in a subdued way. The hills are not so high, the scenery is not so rugged, the soil is not so sterile, the vegetation is not so scanty, and the population is not so sparse.

Fig. 22. Typical Highland Scenery. Loch Long and Loch Goil.

We find then in Scotland **three different kinds of hills** and **three distinct types of scenery** which are characteristic respectively of the Highlands, the Southern Uplands, and the Lowlands. The Highland peaks rise into bare rock, sometimes rugged, splintered, and pinnacled, sometimes upheaving a huge, rounded shoulder of rock terminating in a stupendous precipice. In the Southern Uplands the wildness, ruggedness, and grandeur of the Highlands are softened and modified. The scenery is beautiful, but the outlines of the hills are smoother and more rounded. Yet there is a pure and softly flowing sweep of contour and a charm of delicate colour about these green and treeless summits, found nowhere else in Scotland. The Lowland hills are tablelands, with undulating surfaces, rising into no prominent peaks, and thus differing from both the other types. As a rule the sides of these hills rise very steeply from the low ground in high escarpments, but once the top is gained

Fig. 23. The Lowther Hills with village of Leadhills in foreground. Typical scenery of the Southern Uplands.

one can walk for many miles over bare moorland, the surface rising or falling within the limits of a hundred feet or so.

The fundamental features of the build of Scotland have been indicated in the foregoing paragraphs. There are two other rock systems, which, although of less importance, deserve brief mention. The Old Red Sandstone series immediately underlies the Carboniferous System. As the name implies, these rocks are mainly composed of old sandstones. They form the north-west and also the south-east border of the Lowlands, for a continuous strip of these

THE BUILD OF SCOTLAND

rocks runs right across Scotland on the Lowland side of the Highland Boundary Fault, and another strip (almost unbroken) bounds the Lowlands on the south-east. Another outcrop of Old Red Sandstone is found round the shores of the Moray Firth, widening in the north to include nearly the whole of Caithness. This strip is of great importance, for it is much softer than the Highland schists, and so forms a broad, lowland border on the eastern side of the Highlands. Without exception

Fig. 24. Scenery of the Central Lowlands. The Kilbarchan Hills in the background illustrate the "plateau" type of hills.

every one of the towns and larger villages of the North-West Highlands is situated on the eastern coastal strip. We have here therefore a very striking example of the dependence of the distribution of population on the nature of the rocks.

Some of the islands of the Inner Hebrides, notably Skye and Mull, are formed, not of Highland schists and gneisses, but of huge, tabular blocks of volcanic rock. The volcanoes that poured forth these great sheets of lava were active in comparatively recent times—geologically speaking, of course, for your geologist deals only in millions of years. Skye, Mull, and several adjoining islands once formed part of a continuous sheet of lava that stretched to the north of Ireland. These rocks are the youngest in Scotland. In the south of Skye the tabular sheets of lava are pierced by an allied igneous rock called gabbro. This rock is black, hard, and coarsely crystalline, and it gives rise to the wildest, sternest, and most rugged scenery in all the British Isles. The Cuchullin (or Coolin) Hills of Skye are the paradise of the British rock climber.

The **dominant structural lines of Scotland run from north-east to south-west**. The directions of the Southern Upland Boundary Fault, the Highland Boundary Fault, and the Glen More Fault bear witness to this fact. Many minor structural features run from north-west to south-east, that is, in a direction at right angles to the first. A glance at the north-west of Scotland will show that the sea-lochs of that coast almost invariably take one of these two directions. The **Minch** between the Outer Hebrides and the mainland also runs from north-east to south-west. The sea floor of the Minch sinks to great depths. It is much farther below sea-level than most parts of the Continental Shelf. The origin of the Minch seems to be similar to that of the Central Lowlands. The Outer Hebrides were once joined to the mainland, but the country was cracked by two parallel faults, between which a great block of rock slowly sank till it was carried far below sea-level. The only difference is that the foundering of the Central Lowlands did not proceed so far. Both areas are true "rift valleys" similar in every way to the classic example, the Great Rift Valley of Africa. The sinking of the Minch and the development of lines of structural weakness are probably connected with the foundering of the ancient continent (to which is given

THE BUILD OF SCOTLAND

the name of Arctis) that once joined Scotland and Scandinavia, and stretched north-west as far as Greenland.

Fig. 25. Cuchullin Hills, Skye.

CHAPTER VIII

THE RIVERS OF SCOTLAND

The principal rivers of Scotland are the **Tay**, the **Forth**, the **Tweed**, and the **Clyde**. Of less importance yet still worthy of note are the Spey, the Dee, the Annan, and the Nith. The other rivers are mainly of local interest. Sufficient information regarding them can be obtained from the map. Of the four great rivers, three flow from north-west to south-east, and the Clyde, which flows in an opposite direction, is only an apparent exception to the general rule, for, as we shall see later, there is evidence to show that it once followed a similar course to the others. The dominant direction, then, is from north-west to south-east, at right angles to the "grain" or line of outcrop of the rocks. We conclude that the surface of Scotland in former times had a tilt from north-west downwards to south-east, and this conclusion is borne out by other evidence.

Since the north-west to south-east course of the main rivers of Scotland originated in *consequence* of the tilt of the country these rivers are called **consequent rivers**. These rivers developed tributaries, many of which found it easy to excavate a channel along a soft bed of rock, that is, parallel to the grain or outcrop of the rocks. Since these streams developed later they are called **subsequent rivers**. The River Tay immediately above and below Loch Tay is a subsequent river. These terms are very convenient, as will be seen in later paragraphs.

We are apt to consider rivers, and indeed the whole scenery, of a country as fixed and immovable. But this is very far from being true. If, as Huxley said, our pulses beat centuries instead of seconds, we should see the whole character of the country slowly changing before our eyes. Rivers are continually at war, endeavouring to increase their own territories at the expense of their neighbours. There is a constant struggle for existence going on among them in which the law of the "survival of the fittest" is as inexorable as among plants and animals. When a river steals from another it is guilty of "piracy," and a river that has its head-waters diverted to another basin is said to be "beheaded." Let us look at one or two instructive examples in Scotland.

In the heart of the Cairngorm Mountains there is a loch called Loch Avon. From this lake flows the River Avon in an east-north-east direction (see Fig. 26). All at once the stream turns sharply to the left and flows at right angles to its former course, in due time entering the Spey. This sudden change of direction is in itself significant. But note that the head-waters of the Avon are exactly in line with the head-waters of the Don, and only half a mile distant. Again, the valleys of the Upper Don and the Upper Avon form one continuous valley without any break. All the evidence points to the view that the River Don once flowed from Loch Avon. The valley of the Avon, however, was eroded more vigorously than the

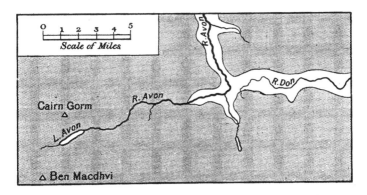

Fig. 26. Sketch-map to illustrate "piracy" by the River Avon. The shaded area is more than 1500 feet above sea-level.

Don valley, and the head of the former river was pushed southwards until it diverted the Upper Don into its own valley. There has been piracy on the part of the Avon, and the Don is now a beheaded river.

Consider next the case of the **Clyde**. Fig. 27 shows that the Gare Loch and Loch Goil are exactly in one line. In addition the narrow neck of land between the two lochs is trenched by a deep pass leading from the head of the Gare Loch to the shores of Loch Long opposite the mouth of Loch Goil. It seems likely that Loch Goil and the

Gare Loch were continuous in former times. Again, the drainage of the mountainous area between Loch Long and Loch Lomond is very peculiar. Several of the streams rise within a short distance of Loch Long. But they do not take the obvious and easy route into the loch. They deliberately, as it were, turn their backs on Loch Long and carve valleys right through the heart of the hills in order to reach Loch Lomond. There are several other peculiarities in this region, but these will suffice. What is the explanation? All the rivers of this area once rose farther west and flowed in consequent courses towards the south-east. These were the original head-waters of the Clyde. But the later formation of Loch Lomond and Loch Long beheaded these streams and diverted the waters to the Atlantic. Then in the dry valley that was left near Greenock by the diversion of these rivers to the west, a short river developed, draining *north-west* into the newly-formed river or loch. This rapidly occupied much of the old valley of the Clyde, with the curious result that the direction of flow was reversed. Such a river is called an **obsequent river**, and the Clyde is the best example in Scotland.

Let us consider next the **Nith**. This river rises in the Southern Uplands and flows north-west to the border of the Ayrshire Plain, whence the route to the Firth of Clyde seems the natural one. But the Nith swings round in the opposite direction, and then trenches a long valley right through the heart of the Southern Uplands. It is obvious that if the relief of the land had been the same formerly as now this course would have been impossible. We must imagine the Central Lowlands not yet worn down to a plain. The river must have originated on a plateau that had a gentle tilt to the south-east. Its source lay much farther to the north-west, perhaps even in a spot now foundered beneath the waves of the Atlantic. Thus we

THE RIVERS OF SCOTLAND 55

are continually brought face to face with the conception of Britain extending in past ages much farther west—a conception that is fundamental to the proper understanding of many problems in British geography.

In length and volume the **Spey** is surpassed only by the Tay in Scottish rivers. It is clearly a subsequent river, having eroded its valley parallel to the grain of the rocks. Its long straight valley is utilised by the Highland Railway from Perth to Inverness. The river is useless for navigation; its course is too rapid. It is a favourite

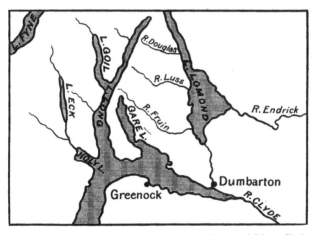

Fig. 27. Sketch-map to illustrate physical history of River Clyde.

haunt of wealthy anglers in the season, for it is one of the best salmon rivers in Scotland. The **Tay** rises in a little corrie far up the slopes of Ben Lui. For the first half of its course it is a subsequent river, in its lower half it is consequent. Notice that the lower consequent course is continuous with the lower Tummel and the Garry. It is obvious that the consequent Garry-Tummel-Tay was the original main river. A subsequent tributary developed the valley in which Loch Tay now lies. It beheaded several of the head streams of the Forth, and at their expense grew larger than the parent consequent river. This subsequent tributary because of its size now arrogates the name Tay, whereas the Tummel and the Garry indicate the original main river. The Pass of Killiecrankie, through which the Garry foams and rages, is worthy of notice. It seems impossible that this narrow and wild defile should contain a river, a road, and a railway. For varied beauty of scenery one might justly claim for the Tay the first place among all the rivers of the British Isles.

The head streams of the **Forth** rise on the slopes of Ben Lomond. In its lower course the Forth meanders in a remarkable manner. The "links of Forth" are noted for their fertility. The chief tributary, the Teith, drains a lake district of wonderful beauty.

The **Tweed** is the principal river of the Southern Uplands. For historical interest combined with beauty of scenery it is the most interesting river in Scotland. As George Borrow said, "Which of the world's streams can Tweed envy, with its beauty and renown?" The tide of border warfare ever flowed strongest in Tweeddale; its banks are crowned by gray ruins of castles, abbeys, and old mansions; many of the border ballads have their scenes placed in this valley, and memories of Scott, of Mungo Park, and the Ettrick Shepherd rise at the name of Tweed. The river is famous for its salmon and trout, indeed, it is quite the most frequented fishing stream in Scotland.

CHAPTER IX

LAKES, LOCHS, AND ISLANDS

THE Southern Uplands contain a number of beautiful lakes, but the most lovely and most famous lakes of Scotland are found in the Highlands. Nearly all the large lakes occur in the valleys of rivers. The river valley has been widened and deepened by some agency, and the

Fig. 28. Loch Katrine.

filling with water of the basin so formed has produced a lake. Most of the Scottish lakes, therefore, are long compared with their width. Some of them are very deep, the lake floors in some instances sinking far below sea-level. The valleys in which they occur in every case show clear evidence of having been occupied by ice, and it is generally believed that the lake basins have been eroded by glaciers. Some of the smaller lakes have been formed by the damming back of water due to moraines left by the glaciers of the Ice Age.

Loch Lomond is the largest and one of the most beautiful of Scottish lakes. Its bottom sinks far below sea-level, and its rocky sides bear the marks of grinding and scoring by vanished glaciers. Quite as lovely, although not so large, is **Loch Katrine**, a lake as useful as it is beautiful, for it furnishes abundant supplies of the purest water to the great city of Glasgow, thirty miles away. These two lakes are most visited by tourists. There are many others, however, that are almost or quite as beautiful, but they are less well known. The mountain pass leading to Loch Katrine and called the **Trossachs** is worthy of mention. Although not the grandest, it is one of the most beautiful mountain passes in Scotland. This district was made famous by Scott's descriptions in *The Lady of the Lake*. **Loch Morar,** in western Inverness, is interesting because of its great depth. The bottom of Loch Morar is the deepest part of the British continental shelf, for it lies more than a thousand feet below sea-level. Only two other lake basins in Europe sink so far below the level of the sea.

The most striking feature of the west coast of Scotland is the way in which it is cut up by sea lochs that run far into the land. The sea lochs of the Scottish Highlands are long and narrow; they do not decrease in breadth as they are traced farther into the land; in fact, the lochs are often narrowest at the mouth, where, too, there is frequently a shallow bar; the sides are rocky and precipitous, and often have waterfalls tumbling down them; the bottom of the loch is often hollowed out into several deep basins. In all these respects the Scottish **lochs resemble the fiords** of Norway; indeed they are often called fiords (compare Fig. 29 and Fig. 76).

These fiords were originally river valleys, which by the

LAKES, LOCHS, AND ISLANDS 59

sinking of the land have been submerged. They are therefore sometimes called "drowned river valleys." But they differ in many respects from the "rias" of south-west Ireland, which are normal, drowned river valleys. The **abnormal features** of the Scottish fiords (and in fact of all true fiords) are their length, narrowness, and straightness, the fact that they run generally in regular directions either from south-west to north-east or at right angles, their deepening *inland*, and the presence of deep rock basins. The origin of the Scottish fiords is probably connected with the fracturing of the west coast due to the foundering of land to the north-west. The newly-formed west coast would be much broken by faults, and along these fault

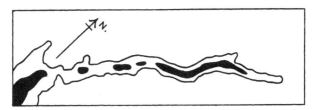

Fig. 29. Loch Fyne, a typical Scottish fiord. The black areas represent deep basins.

lines peculiar valleys were formed, which by submergence gave rise to fiords. Some of the peculiarities of the fiords are almost certainly due to the fact that they were heavily glaciated during the Great Ice Age. The rock basins may be attributed to this cause. The sea lochs of the Firth of Clyde are visited annually by thousands of tourists. Some of the most charming scenery in Britain can be found there. **Loch Long** and **Loch Fyne** are good examples in the Clyde area of typical fiords.

Most of the islands of Scotland lie off the west coast. These islands may be divided into two main groups. The **Outer Hebrides** lie farthest west, about 30 to 50 miles away from the mainland. Close inshore are the **Inner Hebrides**, many of which are separated from the mainland only by narrow channels. North of Scotland, across the stormy waters of the Pentland Firth, are the **Orkney Islands**, and still farther north are the **Shetlands**. In addition to these main groups there are hundreds of rocky islets, many of them dangerous to navigation.

In considering the origin of the islands off the west coast of Scotland it will help us if we look at a particular area, and see how islands might actually be formed from it. In Fig. 30 the black parts represent land more than 500 feet above sea-level. If the sea were to rise or the land to sink 500 feet, it is plain that the land in the figure would be converted into a number of islands separated by narrow straits. These islands are similar in character to those that actually exist round the Scottish coast, and so we conclude that the islands of western Scotland have been produced by a submergence of the land. At the same time we must not forget that the fracturing of the old continent of Arctis, with the development of a number of lines of weakness, first cut up the west of Scotland by numerous valleys which were afterwards invaded by the sea. Just as in the case of the fiords, then, we find that fracturing followed by subsidence best explains the characteristic features of the coast of western Scotland.

The **Shetland Islands** comprise over a hundred separate islands, but only about thirty are inhabited. In rock structure they resemble the Highlands, being composed mainly of schists with fringes of Old Red Sandstone. Many of the islands face the sea in bold cliffs, but they do not rise into prominent hills. There is, therefore, little shelter from the wind, which beats upon these barren islands with wellnigh continuous fury. Trees are unknown, for the violent gales permit the growth of nothing

LAKES, LOCHS, AND ISLANDS 61

more than stunted bushes. At midnight in the summer there is light enough for photographs to be taken and books to be read.

The winds and the rain make it difficult to grow crops in the Shetlands. The fields of oats, barley, and potatoes often blacken and rot before the crop is ripe. Fishing

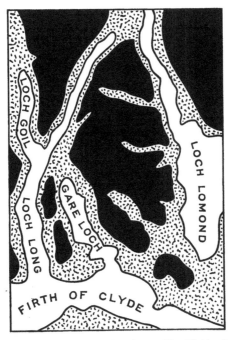

Fig. 30. Sketch-map of Dumbartonshire Highlands to illustrate formation of islands and sea lochs.

is the chief industry. Most of the men are fishermen in the first place, and also farm a small croft. Many of them are skilful cragsmen, and make difficult and dangerous ascents of the cliffs for sea-birds and their eggs. The women knit articles from the fine worsted made from the wool of the native sheep. The little Shetland ponies had need be strong and hardy, for they run in half-wild herds over the islands, picking up a scanty living as best they may. The only town in Shetland is **Lerwick**, which has nearly 5000 inhabitants. It is the centre of the fishing industry.

The **Orkney Islands** are really detached and broken fragments of the Old Red Sandstone plateau of Caithness. They are separated from the mainland by the tempest-beaten waters of the Pentland Firth, where at times the tides rush with such fury as to bring a steamer to a standstill. The Orkneys resemble the Shetlands in many ways: the fine cliff scenery, the flat treeless surface of the islands, the fearful storms that rage on the coasts, the numbers of sea birds, and the occupations of the people. Most of the trade of the group centres at **Kirkwall**, a little town of nearly 4000 inhabitants.

The **Outer Hebrides** consist of a long chain of islands extending for over a hundred miles. So closely connected are the islands of the group that the whole chain is frequently referred to as Long Island. Part of the most northerly and largest island is called Lewis, while the southern part of the same island is called Harris. The names of the other islands can be obtained from the map. The rocks of which the Outer Hebrides are composed are the oldest in Britain. They are coarse gneisses, in appearance not unlike granite, but generally with a banded or streaky structure. The Hebridean Gneiss gives a characteristic type of scenery. On the whole the hills are low and rounded with little of the grandeur of the Highlands or the Inner Hebrides. Their smooth, low domes lie gray and bare, with neither trees nor grass to cover them. The cliff scenery, too, is generally tame. One very characteristic feature is the immense number of lakes dotted over the surface of the bare rock. This is well shown in Fig. 31. Most of these lakes are probably

LAKES, LOCHS, AND ISLANDS 63

true rock basins hollowed out by ice during the Glacial Epoch.

The inhabitants of the Outer Hebrides are engaged in one long, desperate struggle against poverty. It is not too much to say that there are more discomfort and suffering through poverty in these islands than in any other part of the British Isles. The crofters live in poor stone huts, sometimes without windows and chimneys. The smoke from the peat fire in the middle of the room

Fig. 31. Part of Island of Lewis showing remarkable number of small lakes. (After Geikie.)

finds its way out through an opening in the thatched roof. If a person is ill it may take days before a doctor can be obtained. Much of the severe manual labour of the crofts is done by women. The men are fishermen, but let the east coast fishermen carry off all the finest hauls. Many of the women are engaged in making woollen goods such as blankets. In a few places the famous Harris tweeds are made by hand. Even the dyes for the cloth are obtained from the mosses and lichens growing on the islands. **Stornoway**, in Lewis, is the only town. It is the headquarters of the fishing fleets, and in the herring season the bay is crowded with fishing boats from every part of Britain.

The **Inner Hebrides** differ in many respects from the Outer Hebrides. The volcanic rocks of Skye and Mull introduce a different and strong note into the scenery. They occur sometimes as great tables of black rock dropping abruptly to the sea in mighty cliffs. Most striking of all are the gabbro rocks of the Cuchullin Hills, the highest summits of which are over 3000 feet above the sea. They have been carved out of very dark coloured rock, their crests are jagged, splintered, and saw-toothed, and they give an impression of stern, dark, savage grandeur unequalled in Britain. Jura and Islay are composed of schists. The hills of the former rise into graceful cones of a considerable height, but the latter island is fairly flat.

In their mode of life the inhabitants of the Inner Hebrides resemble the people of the Outer Isles, although the conditions are not quite so trying. In addition to the industries formerly mentioned there are distilleries in Islay, Skye, and Mull. In Skye Portree is the chief place, but it is merely a large village. It is a little seaport dealing mainly in fish, sheep, and cattle. Tobermory is a little town in the north of Mull. It is one of the centres of the herring fisheries.

CHAPTER X

VOLCANOES AND GLACIERS

VOLCANOES and glaciers have played such an important part in the physical history of the British Isles that it is advisable to examine their work in some detail. Igneous rocks are divided into two classes, volcanic and plutonic. The **volcanic rocks** cooled on the surface of the earth. They were poured out of volcanic craters or fissures as molten lava, or else were ejected by explosions as volcanic "ash." In both cases a hard, tough rock has been formed that is resistant to the action of the weather, and therefore generally appears as hilly ground. The highest parts of Wales, the Lake District, the hills of the Scottish Lowlands, the hills of Mull, the high plateau of Antrim, have all been carved out of volcanic rock that gushed out as molten lava or shot into the air as ash.

Plutonic rocks solidified below the earth's crust, generally very slowly and under great pressure. They probably represent the subterranean reservoirs which supplied the fiery material that was ejected at the surface of the earth. By the denudation of the overlying rock these plutonic masses are now exposed at the surface, where they appear as granite or similar rocks. The granite bosses of the Lake District, the Southern Uplands, the Highlands, the Mourne Mountains, the Wicklow Mountains, the Cuchullin Hills gabbro, these are examples of plutonic rocks. These rocks generally give rise to a big, massive, rounded type of hills, unless as in Arran and Skye they have been carved by ice into pinnacled and serrated ridges.

In the lowlands of Scotland one frequently encounters a steep, rocky hill that rises suddenly from the surrounding low ground. These hills are not very high as a rule, but they are so steep and craggy that they form very striking features of the scenery. Edinburgh Rock, Dumbarton Rock, and Stirling Rock are good examples. It is not a coincidence that these rocks bear the names of three of the towns most memorable in Scottish history. They were easily fortified, and therefore in early times the strong castles on these hills became centres of important towns. Such hills are called **volcanic necks**, and mark the places

Fig. 32. Dumbarton Rock. A volcanic "neck."

where the lava and other material rose through the crust of the earth. The throat of the volcano became filled with hard rock, which is not so easily worn away as the surrounding softer rocks and so forms a steep, rocky hill. It is strange to think that the towns of Edinburgh, Stirling, and Dumbarton owe their origin to the fact that in long past ages volcanoes happened to break out in the places where the towns now stand.

The igneous rocks that occur in Skye, Mull, and many

VOLCANOES AND GLACIERS

other smaller islands of the Inner Hebrides are of comparatively recent origin. Rocks exactly similar in character and age are found in north-eastern Ireland. There is little doubt that these rocks once formed a continuous plateau connecting Scotland and Ireland. The foundering of part of western Britain, however, carried the central part of this plateau far below sea-level, and fractured the remaining parts into several islands.

The scenery of Great Britain owes a great deal of its present appearance to the **action of ice.** When early man

Fig. 33. Rock surface near Loch Doon showing characteristic rounding and smoothing by ice.

first crossed to Britain from the Continent the climate of this country was much colder than it is now. The Great Ice Age was just passing away. When it was at its maximum most of the country was covered with a thick sheet of snow and ice, through which the higher peaks projected in sharp, rocky ridges. Great glaciers moved from the higher ground down the valleys to the sea, leaving in their passage unmistakable marks to indicate their former presence long after they had disappeared.

The fact that this country was formerly heavily glaciated

is shown in various ways. In certain places the rocks are rounded and smoothed in a way that is only effected by ice. On this rounded surface there are generally long striations or scratches made by the ice dragging stones over the rock. When a glacier melts it leaves at its sides and its end accumulations of rock rubbish called moraines. Such moraines are found in many parts of this country. Most widely spread of all glacial deposits is the boulder-clay, a tough clay with stones imbedded in it. A large proportion of the surface soil of this country is not derived

Fig. 34. Loch Tay, an ice-eroded basin. In the foreground is seen the delta of the Rivers Lochay and Dochart.

from the weathering of the underlying rocks, but consists of boulder-clay of glacial origin.

Many of the most characteristic features of the scenery of Britain are due to ice. Much of the charm of our most attractive districts is due to the number and beauty of the lakes that are found there. Nearly all the lakes in the British Isles (except the shallow loughs of the Irish Plain) have been caused by ice action in some form. In some cases, as at Killarney, the drainage has been dammed by

VOLCANOES AND GLACIERS

morainic material, more frequently rock-basins have been excavated in valleys by glaciers. These rock-basins are generally fairly deep, and not infrequently sink far below sea-level. The deep basins of many of the sea lochs of Scotland are probably due to the same cause.

Much of the ruggedness and wildness of the scenery of the mountainous parts of the British Isles is due to sculpturing by ice. For example, consider Fig. 35, which shows the summit of Braeriach, the third highest mountain in the British Isles. Between the three summits, over

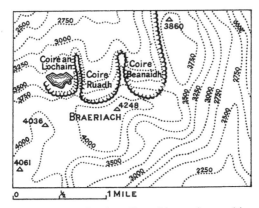

Fig. 35. Map showing effect of ice-erosion resulting in formation of corries.

4000 feet in height, it is plain from the contour lines that there extends a fairly broad, flat-topped area, in fact a small tableland. This level area is suddenly trenched to the north by three semi-circular precipices forming the heads of the three coires, Coire an Lochain, Coire Ruadh, and Coire Beanaidh. These three coires are separated by sharp arêtes, narrow knife-edged spurs that run north from the main mass of the mountain. These deep coires, walled in at their heads by stupendous precipices, and their boundary ridges form by far the most impressive part of the scenery of the Cairngorm Mountains. The northerly direction in which the coires face is significant, for in all probability these great hollows have been excavated by ice. The same features can be seen on Snowdon. Before the Ice Age this mountain was probably round-topped, with easy slopes on all sides. The wonderful cwms[1] and narrow serrated spurs of Snowdon have been sculptured by ice. In the Lake District exactly the same features are seen on the east flank of Helvellyn round Red Tarn.

In summing up, then, we may say that fire and frost have provided some of the most picturesque scenery in Britain. Volcanic action supplied enormous blocks of rock which were sculptured into their present forms by various agencies of which one of the most important was ice.

CHAPTER XI

THE HIGHLANDS AND THE NORTH-EASTERN LOWLANDS

THE Scottish Highlands afford an excellent example of that type of mountainous country known as a **dissected plateau**. Of mountain ranges in the true sense of the word there is none. Looking from the Lowlands towards the Highland line one seems to see a long range of mountains crossing the country, but this is merely the rough, dissected edge of the Grampian plateau. If the rock removed from the Highland valleys by rivers and glaciers were replaced, the land would be restored to its original condition of an undulating tableland. The direction of the consequent

---

[1] Coire (sometimes spelled corrie) is pronounced korry, cwm is pronounced koom: these curious hollows are known also as cirques.

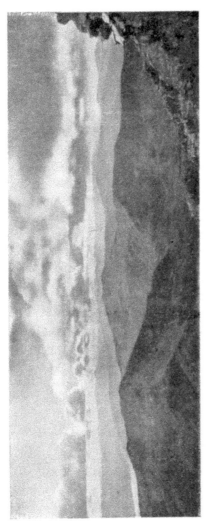

Fig. 36. View from summit of Ben Lawers. The photograph illustrates the fact that the Scottish Highlands form a dissected plateau.

rivers of Scotland indicates that this plateau had a slight tilt to the south-east. Towards the north-west the plateau was fractured by the earth movements that separated it from the old continent of Arctis. Towards the south-east the tableland of schists and gneisses is cut off sharply by the Highland Boundary Fault.

The Highlands are divided into two parts by the long straight valley of Glen More, which marks an important fault. North-west of Glen More are the North-West Highlands, south-east of the fault are the Grampian Highlands. All the highest peaks are in the Grampian Highlands. There are half a dozen over 4000 feet in height. **Ben Nevis** is the highest single peak. The highest continuous area is found among the **Cairngorm Mountains** on the borders of Aberdeen, Inverness, and Banff. Here within a few miles radius are found Ben Macdhui, Braeriach, Cairntoul, and Cairngorm, all over 4000 feet high. Since the rainfall of the West Highlands is greater, erosion has been more active there. The peaks are more pointed, the ridges are sharper and more serrated, the massive, rounded mountain summits characteristic of the Cairngorm Mountains are much less frequently seen. In the west the original plateau has been more deeply dissected, or (in the language of physical geography) denudation in the west has reached a more mature stage. The fault of Glenmore afforded a line of weakness to the agents of erosion which have etched along it a deep valley right across Scotland. This valley is occupied by three narrow lakes, and it was therefore not difficult to construct the **Caledonian Canal**, which it was hoped would bring trade to this part of the Highlands. But the surrounding district is sparsely populated, and the canal does not open communication between any large centres of trade, and so it is now almost entirely a route for tourists.

In almost every phase of his industrial life the

THE NORTH-EASTERN LOWLANDS 73

Highlander is struggling under a heavy handicap. The climate is wet and sunless, the soil is sterile, communication is exceedingly difficult, and useful minerals are practically absent. Agriculture is confined to the valleys, and even there it is almost impossible to grow wheat. If the crofter's patch of oats ripens, and his potatoes are unblighted, he has had a successful season. The chief industry is sheep-rearing, which is carried on over all the Highlands. Argyllshire has more sheep than any other county in Scotland, although not so many *in proportion to size* as some of the counties of the Southern Uplands. Wide areas of the Scottish Highlands are fit only for grouse-moors and deer-forests. It should be noted that one might walk for miles over the so-called forests and never see a tree. In summer and autumn the charms of sport and the magnificent scenery of the Scottish Highlands attract thousands of holiday makers from the south, and catering for these tourists is now one of the chief occupations of many parts of the Highlands. If communication were easier more use might be made of the waterfalls for driving machinery. On a stream running into Loch Ness are the Falls of Foyers, which supply the power to make electricity which is used for smelting aluminium in electric furnaces. At the head of Loch Leven (off Loch Linnhe) is the village of Kinlochleven, where electric plant has also been erected for smelting aluminium. The power is obtained from an artificial fall. There are quite a number of whisky distilleries in various parts of the Highlands. **Campbeltown** on the Firth of Clyde is noted for **whisky**, and in Islay and other districts there are also distilleries.

Inverness is a "nodal" town, that is, it is situated at a node or crossing place of routes. It stands at the end of the Caledonian Canal; the traffic along the North-eastern Lowlands must pass through the town; and an important route from the Central Lowlands through the heart of the

Grampian Highlands emerges on low ground near Inverness. In former times it was of great military importance because it commanded so many routes through the Highlands, but nowadays it is known chiefly as a tourist centre. If the population of the Highlands were not so small Inverness would certainly be a large and important town, because of its situation at the crossing of such easy routes through a difficult country. A few miles east of the town is the moor where Culloden was fought in 1746.

At the other end of the Caledonian Canal is **Fort William**, formerly another important military station. Like Inverness it is a busy place in summer, and an excursion centre by road, rail, and steamer. Ten miles lower down Loch Linnhe is the entrance to Loch Leven. Opening into this loch is lonely Glen Coe, one of the sternest and most desolate glens in Scotland. Memories of the terrible massacre of 1692 still seem to haunt this wild and lonely place. At the entrance to Loch Linnhe is **Oban**, the chief centre of the steamer traffic of the West Highlands. Until 1894, when the West Highland Railway was opened, the only communication between the busy towns of the Clyde and most of the West Highlands was by boat.

Aberdeen may be considered either in the Highlands or in the North-eastern Lowlands. The neighbouring rocks are Highland schists with granite intrusions, but beginning at Aberdeen there stretches northward the broad plain of Buchan, continued into the Moray Plain on both sides of the Moray Firth. These lowlands are fertile districts, where agriculture is important. Cattle-rearing is one of the chief branches. Aberdeen is by far the largest city north of the Highland line. It is situated where the Highlands come close to the sea, forcing all traffic along the coast to pass near the town. Just where this narrow passage is crossed by the largest river of the district

THE NORTH-EASTERN LOWLANDS 75

Aberdeen is situated. The great bosses of grey granite in the neighbourhood are quarried, and the stone shipped for monumental work, bridges, docks, and similar massive structures. Aberdeen is the chief fishing station on the east coast of Scotland. Many of the fishing boats are built and engined in the town.

Nearly all Caithness and most of the coastal plain stretching from the north of Scotland round the Moray Firth are composed, not of schists, but of Old Red Sandstone. These lowlands offer many strong contrasts to the Highlands that they fringe. They form a plain or low tableland with no prominent mountains. The climate is much drier and sunnier than that of the land farther west. This obviously results from the fact that they lie in the lee of the Highlands, or, to use an expressive phrase, in the "rain-shadow" cast by the Highlands. Gaelic is not spoken in the North-eastern Lowlands. The predominating influence here has been Scandinavian, as may be seen from the place names. Fishing is the principal industry, **Wick** and **Thurso** being the chief stations. Some of the Old Red Sandstone rocks split easily into large, flat slabs, which furnish the best paving stones in this country.

The eastern coastal strip offers the easiest route for a railway, and so we find the North-eastern Lowlands traversed by a railway line that goes right to Thurso on the north coast. Between Aberdeen and Elgin this railway is called the Great North of Scotland Railway. Beyond Elgin the line is called the Highland Railway. From Dingwall a branch goes west to Loch Alsh opposite Skye, the only railway that crosses the Highlands from east to west. The main Highland Railway leaves Perth and attacks the Grampians by means of the Tay valley, and then by the tributary valleys of the Tummel and the Garry. The carrying of the line through the narrow Pass of Killiecrankie was a wonderful feat of engineering. The summit level of 1500 feet is reached at the Pass of Drumochter between the Garry and the Spey. The line next uses the valley of the latter river, and then swings sharply to the left, and reaches Inverness where it joins the Elgin

Fig. 37. Pass of Killiecrankie.

branch already mentioned. The West Highland line runs from Glasgow to Fort William, and thence to Mallaig opposite Skye. It keeps just far enough inland to avoid the fiords of the west coast.

CHAPTER XII

THE SOUTHERN UPLANDS

On the north-west the Southern Uplands are sharply separated from the Central Lowlands by the Southern Upland Boundary Fault, which runs in a fairly straight line from Dunbar in Haddington to Girvan in Ayrshire. On the south-east the Southern Uplands are bounded by the coastal plain of the Solway Firth and the plain of lower Tweeddale. Between the Solway and the Tweed, however, the hilly ground continues over the Border as the **Cheviot Hills**, which join the Southern Uplands to the Pennine Uplands of England. Like the Highlands, the Southern Uplands form a dissected plateau, the surface of which was somewhat lower than that of the Highland plateau, for, as we have already seen, the original tableland had a tilt to the south-east. This is confirmed by the heights of the mountains, which in the Southern Uplands do not reach 3000 feet. Merrick (2764 feet), on the borders of Ayrshire and Kirkcudbright, is the highest. Seen from the Central Lowlands the scarped edge of the tableland presents the appearance of a range of hills, and on this belief names have been given to the plateau edge. The north-eastern escarpment of the Southern Uplands is called in one part the **Moorfoot Hills** and elsewhere the **Lammermoor Hills**. They are separated by the Gala Water, which cuts a deep notch into the tableland.

The scenery of the Southern Uplands is generally beautiful. Many parts are wild, stern, and lonely, but there is not the same ruggedness as in the Highlands (see Fig. 23). This results partly from the fact that the rocks are not so hard and resistant as the Highland schists, and partly because the original tableland was not so high. The rocks are largely of Silurian age. Along the boundary fault they abut against the much younger Carboniferous and Old Red Sandstone rocks of the Central Lowlands. In Kirkcudbright there are three large intrusions of granite, and the nucleus of the Cheviot Hills is also igneous, being composed of volcanic rock that was poured out as molten lava in Old Red Sandstone times.

Except for some unimportant seams in Nithsdale there is **no coal** in the Southern Uplands. Consequently we do not find any very large manufacturing towns in this part of Scotland. In the Lowther Hills, on the borders of Lanark and Dumfries, **lead-mining** has been carried on for hundreds of years, and the little villages of Leadhills (Lanark) and Wanlockhead (Dumfries) are the only lead-mining places in Scotland at the present time. These villages are 1400 feet above sea-level, and claim to be the highest villages in the British Isles. In Dumfriesshire there are valuable quarries of **red sandstone**, which is much in demand for building purposes. One of the exposures of **granite** in Kirkcudbright touches the Solway Firth, and is thus in a favourable position for quarrying and exporting. **Dalbeattie** is the centre of the granite industry of the south.

The broad valleys and coastal plains of the Southern Uplands are well suited for agriculture. There is a significant **contrast between** the farms of the **west** and those of the **east**. Wigtownshire is noted for its dairy produce, while in the lowlands of Berwickshire grain is grown fully as successfully as in the more favoured districts of East Anglia. The rich grass pastures of the west and the sunny wheat lands of the east are alike the result of geographical causes, the controlling factor in this case being climate. **Sheep-rearing** is the principal industry of the Southern Uplands. All the counties in the south of Scotland possess

THE SOUTHERN UPLANDS

large numbers of sheep, in fact Roxburgh has more sheep in proportion to its size than any other county in the British Isles. It is therefore not surprising to find that the chief manufacture of the Southern Uplands is the making of **woollen goods**. The principal seats of this industry are in the basin of the Tweed. Supplies of raw material are easily obtained, and in former times the water-power of the streams was used to drive machinery. Since the application of steam to manufacturing processes the industry has thriven, but not nearly so much as in centres like Yorkshire, where abundant coal is at hand.

Hawick on the Teviot is the largest town in the Tweed basin, and is the chief place in Scotland for the making of hosiery. Tweed[1] cloth is also made, but the principal centre for the manufacture of this material is **Galashiels** on the Gala. Down the river from Galashiels is **Abbotsford**, the home of Sir Walter Scott. Another famous building in this locality is **Melrose Abbey**, in some respects the most beautiful ruin in Scotland. **Peebles** is also engaged in the woollen trade. It has a beautiful situation on the Tweed, which made it in former times a favourite place of residence of many Scottish kings. It is still a much frequented inland watering place. Among the towns of the Southern Uplands **Dumfries** ranks next to Hawick in size. It is situated on the Nith at the lowest point at which the river can be bridged conveniently. It has several woollen mills, but its importance arises chiefly from the fact that it is the principal **market and commercial centre** of the south-western counties. Several railways converge at Dumfries. The main line of the Glasgow and South-Western Railway from Glasgow to Carlisle runs through the town. Another important line goes from Dumfries to Stranraer. It carries a considerable number

---

[1] The word "tweed" has probably no connection with the river but is a corruption of "tweel," the local name of twilled cloth.

of English passengers who wish to take advantage of the **Stranraer to Larne route** across the Irish Sea, the **shortest route** between Great Britain and Ireland.

It is obvious that the Southern Uplands present a formidable barrier to traffic between the busy towns of the Central Lowlands and all parts of England. Luckily, however, the hill mass is trenched completely through by more than one valley. We saw in a former chapter that the Nith, after emerging on the lowland plain, turns back and cuts a valley through the whole width of the Southern Uplands to the Solway Firth. This valley is used by the **Glasgow and South-Western Railway,** the summit level of which is about 700 feet above sea-level. The line enters the Southern Uplands at New Cumnock, and emerges on the coastal plain of the Solway at Dumfries. Thence it runs through Annan to Carlisle, where it connects with the Midland Railway. An examination of the map reveals that a little east of Nithsdale the Southern Uplands are almost completely cut through. The Clyde flowing north and the Annan flowing south approach each other very closely, and at their nearest point the valleys are connected by a pass. This is the route used by the **Caledonian Railway** from Glasgow to Carlisle. The highest point of the pass between the Clyde and the Annan is Beattock Summit, which is 1000 feet above the sea. The Edinburgh trains by this route join the line at Carstairs. At Carlisle the Caledonian Railway connects with the London and North-Western Railway. The valley of the Gala-Tweed also trenches right across the Southern Uplands, and this route is used by the **North British Railway** from Edinburgh to Carlisle. The line runs down the Gala past Galashiels to the Tweed, then turns south-west up the Teviot, and after leaving Hawick crosses the hills by a difficult route to Liddesdale, whence the way is easy to the Solway coastal plain and Carlisle. From the fact that

THE SOUTHERN UPLANDS

this line traverses the country immortalised by Scott it is often called the Waverley Route.

The three routes we have just described, namely, the Nith route, the Clyde-Annan route, and the Gala-Teviot-Liddel route, all cut through the heart of the Southern Uplands. Two other railways make use of the coastal fringes on the east and on the west of the Uplands. From Edinburgh the North British Railway takes the East Coast

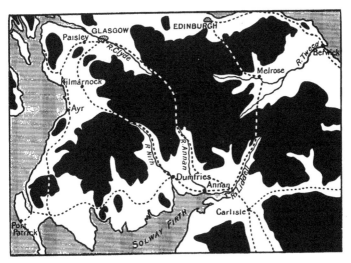

Fig. 38. Routes across the Southern Uplands. The black areas represent land more than 500 feet above sea-level.

Route by Dunbar to Berwick, where it joins the North-Eastern Railway. The Glasgow and South-Western Railway runs along the west coast from Ayr to Stranraer. On the west the Southern Uplands at several points face the sea in cliffs, and the railway is therefore forced inland for a considerable part of its course. This line naturally carries a large share of the Irish passenger and mail traffic.

If the Southern Uplands must yield to the Highlands

in beauty of scenery and charm of colour, yet in historical associations they are infinitely richer. The scenes of most of the Border ballads are placed in the Southern Uplands. Many of the most famous men Scotland has produced were born in the Southern Uplands, or are intimately associated with this district by means of long residence. Bruce, Hogg the Ettrick Shepherd, Allan Ramsay, Telford the engineer, Mungo Park and Joseph Thomson, Symington the father of steam navigation, Carlyle, Scott, and Burns—these are the names of a few that will ever be associated with the Southern Uplands.

CHAPTER XIII

THE CENTRAL LOWLANDS

WE have already seen that the Central Lowlands are bounded on the north-west by the Highland Boundary Fault and on the south-east by the Southern Upland Boundary Fault. The part of Scotland that lies between these two lines of fracture is in the main of a lowland character, but by no means altogether so. There are several hill masses, of which the most important are the **Sidlaw Hills** north of the Tay, the **Ochil Hills** between the Tay and the Forth, the **Campsie Fells** between the Forth and the Clyde, the **Kilbarchan Hills**[1] still farther to the south-west, and the **Pentland Hills** south-west of Edinburgh. These hill masses (except the last named) are arranged in a line roughly parallel to the Highland

[1] These hills between Greenock and Ardrossan have no generally accepted name. The name Kilbarchan Hills was used by Prof. James Geikie many years ago and is a convenient designation for the whole plateau.

THE CENTRAL LOWLANDS

Boundary Fault and about ten miles from it. The belt of low ground between these hills and the Highlands is a fertile tract called **Strathmore**[1].

The Central Lowland Hills are all of the plateau type. They are remnants of great sheets of volcanic rock that once occupied a much larger area. The breaks between the hills have been caused by the rivers Tay, Forth, and Clyde, and these breaks have been of the utmost value for communication. The **Ochil Hills** may be taken as typical. They consist simply of a great slab of solid lava, some twenty miles long, ten miles broad, and 2000 feet high. These lava plateaus are not deeply dissected like the Highland plateau. Their tops are fairly level and afford

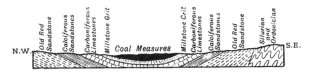

Fig. 39. Section across the Central Lowlands showing how the Coal Measures occupy a basin.

examples of immature topography, none of the valleys being deeply cut. A band of Old Red Sandstone occupies Strathmore and another band of the same age borders the Southern Uplands. The rest of the rocks (except some small patches) are Carboniferous. Of this system the **Coal Measures** are of course the most important. The rocks of the Central Lowlands are arranged in the form of a shallow trough, as is shown in Fig. 39, and the Coal Measures are therefore found near the centre of the trough. The simplicity of the trough structure is marred by an upswelling or anticline of rocks forming the high moors along the Lanark-Ayr and Renfrew-Ayr borders. This anticline,

---

[1] The Great Strath. A strath is a wide, flat valley in contradistinction to a glen.

therefore, separates the Lanarkshire Coalfield from the Ayrshire Coalfield.

The **Clyde valley** is commercially by far the most important part of Scotland. Almost a half of the whole population of the country is settled there. This part of Scotland only rose to predominant importance in the nineteenth century. There were two main causes, first, the development of the mineral wealth of the rich coalfields of the west, and, secondly, the great growth of Atlantic trade. **Glasgow** is the heart of industrial Scotland. It contains a million inhabitants, and is the second largest city in the United Kingdom. The site of Glasgow has a high degree of nodality. The city stands at the lowest point at which the Clyde can be bridged, and, therefore, traffic along the sides of the river estuary coalesces and diverges at Glasgow. The Clyde valley is separated from the fertile Ayrshire plain by the anticline of hills forming the boundary between Renfrewshire and Ayrshire. This hill barrier is cut through by two deep valleys, one at Paisley, another at Barrhead, both of which point towards Glasgow. The easy Annan-Clyde route from England has a natural termination at Glasgow. Scotland is narrowest between the Forth and the Clyde, and the land is fairly flat, so that most of the east and west traffic of Scotland begins or ends at Glasgow. During the last hundred years Glasgow's rapid growth has been largely due to the development of the iron and coal of the Lanarkshire Coalfield. Every branch of the **engineering** industry is carried on. Locomotive building is one of the most important. More locomotives are built in Glasgow than in any other town in Europe. Bridge work is another typical speciality. Those wonderful structures, the Tay Bridge and the Forth Bridge, were erected by a Glasgow firm. **Cotton spinning** is largely carried on in the city, and muslins and curtains are produced on a large scale.

THE CENTRAL LOWLANDS 85

Glasgow takes high rank as a **seaport**, being surpassed as a rule only by four other British ports[1]. Great sums of money have been spent in making the Clyde navigable to Glasgow. A century and a half ago people could wade across the river without wetting their knees, where now great ocean liners sail in safety.

Above Glasgow the Clyde is of no importance as a commercial highway, the railways being the chief means of communication. All the large towns of this district are situated on the Lanarkshire Coalfield, and are engaged in various branches of the coal and iron trade. Before the coalfield was worked there was not a single large town in the district. **Coatbridge** and **Airdrie** stand side by side on the northern part of the coalfield. Coatbridge is the chief centre for the smelting of iron in Scotland, and possesses more than half the blast-furnaces of Lanarkshire. Both towns have big steel works and rolling mills. Boilers, tubes, and heavy metal goods of all kinds are made. **Motherwell**, on the north bank of the Clyde, is one of the chief steel centres in Britain. Before the exploitation of the coalfield there was not even a village where this large town now stands. It makes ingots, rails, girders, boiler-plates, and many other kinds of steel goods. It specialises also in roof and bridge work. On the opposite bank of the Clyde is **Hamilton**. Many of the inhabitants are engaged in the surrounding coalpits. **Wishaw** is another large coal town. Many branches of engineering are also carried on in both these towns. Above the coalfield the Clyde valley completely changes its character. It is a beautiful fruit-growing district, the scenery round the famous Falls of Clyde being especially lovely. **Lanark** is the chief town of this part of the valley. It is an

---

[1] The order varies according as tonnage or value is considered. Manchester and Glasgow struggle for fourth place in the annual returns of value of exports and imports. In the returns of tonnage both rank much lower.

old historic town, with many memories of Sir William Wallace.

From Glasgow to the mouth of the river the banks of the Clyde are lined with **shipbuilding towns** that make this the greatest shipbuilding district in the whole world. The following table gives in order of importance the tonnage for 1912 of the world's most important shipbuilding districts:

	tons
The Clyde	641,000
All Germany	530,000
The Tyne	388,000
All United States	322,000
The Wear	310,000
The Tees	262,000

It is obvious that the Clyde possesses many weighty **advantages for this industry.** The estuary runs into a busy coalfield where iron and steel working and marine engineering are staple industries, and so material and skilled labour are close at hand. The human factor, too, has been no less important. The enterprise and foresight of the citizens of Glasgow in changing a shallow stream into a navigable river made the industry possible. Again, some of the greatest inventions in the history of shipbuilding were made by Clydesdale men.

After Glasgow the chief shipbuilding towns are **Renfrew, Clydebank, Dumbarton, Port Glasgow,** and **Greenock.** The shipbuilding of Renfrew is of a specialised kind. It builds more dredgers than any other town in the world. Clydebank has grown in recent years in the most phenomenal manner. In the last ten years it has increased by 80 per cent., a growth unparalleled by any other large town in the United Kingdom. The rapid increase has been caused by the establishment of large engineering and shipbuilding firms. Among the individual shipbuilding towns Port Glasgow produces the largest tonnage. In

Fig. 40. Shipping on the Clyde.

Greenock sugar refining is still carried on, although the industry was severely shaken by the growth of the Continental beet-sugar industry. **Paisley**, situated seven miles south-west of Glasgow, has the largest **cotton-thread** mills in the world, and contains also many great **engineering** works specialising mainly in machine tools.

The control of economic conditions by geographical situation is well exemplified by the fact that in the Clyde valley below Glasgow there are neither blast furnaces nor steel works. The lower Clyde valley is situated at some little distance from the rich coalfield of Lanarkshire, and therefore the industrial towns are at a disadvantage in crude processes where large quantities must be turned out at small cost. The cost of carriage of coal and ore is too great a handicap. This is a geographical principle of wide application. Whenever the source of power is not in the immediate neighbourhood, we find that the amount of skill and labour expended on the articles produced is large compared with the cost of raw material. The fact that Paisley specialises in costly machine-tools is a good illustration of this principle.

The **Firth of Clyde** can boast of some of the most beautiful scenery in Scotland. The Highland Boundary Fault runs for the most part under the waters of the estuary so that the north-west shores are Highland, and the opposite shores of Renfrew and Ayr are Lowland. The Highland shores are penetrated by long sea lochs, much visited in summer-time by tourists. **Rothesay** in the island of Bute and **Dunoon** on the coast of Argyll are the busiest watering-places. **Arran** is the largest and most beautiful of the islands. In the interior are rugged peaks of granite, deep glens, and ice-worn coires. For the lover of natural history and fine scenery there is no more attractive place in Britain than the island of Arran.

CHAPTER XIV

THE CENTRAL LOWLANDS (*continued*)

IN dealing with the Southern Uplands we saw that Wigtownshire was a dairy-farming county, while parts of Berwickshire were devoted largely to wheat. The same contrast holds between the western and the eastern counties of the Central Lowlands. Fife and the Lothians are the most important grain-growing counties in Scotland, while in Ayr and Renfrew dairy-farming is the chief branch of agriculture. The determining factor is obviously a climatic one. The **Ayrshire Plain** is very typical of the whole of Lowland Scotland. The industries are unusually well-balanced. Agriculture, fishing, mining, and manufacturing are all largely carried on. Dairy-farming and potato growing are the chief branches of agriculture. Part of London's daily milk supply comes from the farms of North Ayrshire. This county produces more iron-ore than any other in Scotland, and its coalfield supplies power to several large manufacturing towns. **Kilmarnock** is the largest. Engineering is the chief industry, but there are also large factories for making woollen goods and boots. **Ayr**, at the mouth of the River Ayr, is a typical county town. It is the market and commercial centre for quite a large agricultural and mining area. There is a considerable shipping trade, a good deal of it being with Ireland. Timber, iron-ore, and limestone are imported, and coal and pig-iron are exported. **Ardrossan**, the other large seaport of Ayrshire, has a similar trade. In addition, however, it is the principal Scottish port for fast passenger traffic with Ireland. Ayrshire is intimately associated with many famous Scotsmen. Wallace, Bruce, and Burns are the most outstanding names.

A glance at the map will show that the hills of the Lowlands: the Sidlaws, the Ochils, the Campsies, and the Kilbarchans, form quite a considerable barrier to communication (see Fig. 41). Obviously the traffic will not pass over the hills but through the gaps between them caused by the Tay, the Forth, and the Clyde. We expect, then, to find these gaps occupied by important towns, and in truth we find at these places three of the oldest and most historic towns in Scotland, namely, **Perth** at the Tay gap, **Stirling** at the Forth gap, and **Dumbarton** at the Clyde gap. The reason for the supreme military importance possessed by these towns in former times is now apparent, for they command the routes through the gaps between the Lowland Hills. All three towns also show us that invariable **evolution of the strategic town of former times into the railway centre of to-day.**

Perth commands routes along the east coast to Dundee and Aberdeen, along Strathmore, and by the Tay valley through the heart of the Highlands to Inverness. It has more historic memories than any other town in Scotland, Edinburgh alone excepted. Its only industries of note are dyeing and dry cleaning. **Stirling** grew up round the steep rock on which was placed the famous Stirling Castle. Its chief industry is the manufacture of woollen goods. Blankets and carpets are made, and it is the principal centre of the making of Highland tartans. **Dumbarton** shares in the shipbuilding and engineering industries of the Clyde.

Just as Glasgow is the chief centre of the manufactures and shipping characteristic of west Scotland, so **Edinburgh** dominates the life of the east. It is only one-third the size of Glasgow, but its historical prestige and the fact that the chief law-courts and government offices are

THE CENTRAL LOWLANDS

situated in the capital, give it an importance out of proportion to its size. Its situation is similar in some ways to that of Stirling. It is a **gap town** standing in the narrow passage between the Pentland Hills and the sea. All traffic along the east coast must therefore pass near Edinburgh; and it must be remembered that until the nineteenth century the route along the eastern coastal plain of Scotland was by far the most important in the country. The exact position of the town in the gap was determined by the steep rock on which a strong castle could be built.

Edinburgh is built on a number of hills, from which fact it sometimes gets the name of Modern Athens. The hills offer many commanding sites for buildings, which have been taken advantage of, and consequently Edinburgh is in many ways the most impressive city in the British Isles. Among the most interesting of the old buildings are Holyrood Palace, St Giles' Church, the Castle, and the house of John Knox. Edinburgh has always been noted for its interest in education. There are more first-class schools in this city than in any other town in Britain. The University has a world-wide reputation for its medical teaching. The Advocates' Library is one of the few libraries which must receive a copy of every book published in Britain. **Paper-making, printing and publishing** are trades that have long been associated with Edinburgh. When the law-courts are sitting an enormous amount of printing must be done requiring much skilled labour, which, on the rising of the law-courts, is set free for the ordinary branches of printing and publishing. The rise of the paper-making industry is due partly to the abundant supplies of pure water from the neighbouring hills. To the same cause must be traced part of the success of the large **breweries** and **distilleries** in the town.

Leith is the port of Edinburgh, and the two towns are now continuous. Among the towns of Scotland Leith takes sixth place, while as a **seaport** it ranks next to Glasgow. Most of its trade is carried on with the continent of Europe, especially the ports of the Baltic and the North Sea. Shipbuilding, engineering, sail-making, rope-making, and other industries with a "salt-water flavour" are carried on in Leith. Higher up the Firth of Forth is the seaport of **Grangemouth**, the trade of which is like that of Leith though on a smaller scale. Timber, ore, flax, and grain are the chief things imported. The town benefits to some extent by its position at the east end of the Forth and Clyde Canal, which cuts across Scotland at its narrowest part. Three miles inland from Grangemouth is **Falkirk**, the chief industrial town of Stirlingshire. **Iron-founding** is the principal industry of the district. Carron, a few miles from Falkirk, has large ironworks which used to be famous for the manufacture of cannon, and indeed gave its name to carronade, a short, smooth-bored gun first made in Carron in the eighteenth century.

The eastern Lowlands south of the Firth of Forth are made up of the counties of Linlithgow, Edinburgh, and Haddington, which together are known as the Lothians. This is one of the best agricultural districts in Scotland. In wheat-growing, for example, although Fife takes the first place, these three counties hold the second, third, and fourth places. Coal is mined in Edinburgh, although the output is less than a third of that of the Lanarkshire field. The most interesting mineral is **oil-shale**, found chiefly in Linlithgow. This mineral when strongly heated gives off various oils from which are obtained petrol, light burning oils, heavy lubricating oils, ammonia, and paraffin wax. In spite of keen foreign competition this typical Scottish industry is in a very flourishing condition. **Bathgate** and

THE CENTRAL LOWLANDS 93

Broxburn are the chief centres of the industry. The town of **Linlithgow** was formerly a favourite place of residence of Scottish sovereigns, and has many interesting historical memories.

CHAPTER XV

THE CENTRAL LOWLANDS (*continued*)

THE peninsula between the Firth of Forth and the Firth of Tay is occupied mainly by the **county of Fife**. Except in the west of the peninsula where the Ochil Hills rise to 2000 feet the land is fairly flat and well suited for farming. Consequently this district is tilled more than any other part of Scotland, and in proportion to its size the county of Fife grows more wheat than any other Scottish county. The district resembles the Ayrshire plain in many respects, for not only is agriculture important, but coal mining is carried on, there are several busy manufacturing towns, fishing is important round the coast, and the villages on the sea coast are popular holiday resorts in summer. The resemblance might be pushed even further, for if Fifeshire is the greatest golfing county in Scotland, then Ayrshire is a good second.

The **Fife Coalfield** has fostered the growth of several manufacturing towns. **Kirkcaldy** is the most important centre in Britain for the manufacture of linoleum and floor-cloth. It is also a busy seaport. In the manufacture of fine linen **Dunfermline** rivals successfully the linen towns of Ulster. Indeed in the making of table-cloths, napkins, and exquisite damask work of all kinds it has no rival in the British Isles. The ruins of a royal palace and a beautiful old abbey remind us of the prominent part Dunfermline once played in Scottish history. **Methil** is situated where the Fife Coalfield touches the sea. It has lately sprung into prominence as the greatest coal-exporting town on the east coast of Scotland. The development of the coalfield is causing some of the towns in this part of Scotland to grow with amazing rapidity. For example, in 1891 Buckhaven had 4000 inhabitants. The 1911 census showed that it had grown to a town with 15,000 inhabitants. **Alloa** in Clackmannan has large breweries and distilleries, and is noted also for the manufacture of hosiery. **St. Andrews** is one of the four university towns of Scotland. Its early importance was due to church influence in pre-Reformation times. Many of the chief struggles of the Reformation took place in St. Andrews. Its university is the oldest in Scotland, but St. Andrews is probably better known as the most famous golfing centre in the world.

The peninsula of Fife is joined to Forfar by the Tay Bridge, which spans the Firth of Tay where it is two miles wide. The bridge leads the railway directly into **Dundee**, the third largest city in Scotland. This town is the chief seat in this country of the manufacture of **jute**. From the coarser varieties sacking and ropes are made, while the finer kinds are woven into curtains and furniture coverings. Millions of brightly coloured jute carpets are sent to Moslem countries to be used as praying carpets. Dundee can import flax cheaply from the Baltic seaports, and has a considerable manufacture of coarse **linen goods**. There are a number of jam-factories that use the fruit grown in the neighbouring Carse of Gowrie. The shipbuilding yards of the Tay turn out about twenty or thirty moderate sized steamers each year. The whaling fleet used to lend some romance to the drab town, but mineral oil has practically killed the whale-fisheries. Farther north on the Forfar coast are the towns of **Arbroath** and **Montrose**, which

THE CENTRAL LOWLANDS

have also manufactures of coarse linen. They are fishing centres, too, like nearly all the little seaport towns of the east coast.

The Central Lowlands are hilly enough to furnish good examples of the **control of routes by the relief of the land**. There are no serious difficulties to be overcome by the railways between Edinburgh and Glasgow. Both the North British and the Caledonian have main lines between these cities. The North British Railway keeps somewhat to the north, and on the whole has a distinctly easier route. The east coast of the Central Lowlands is served chiefly

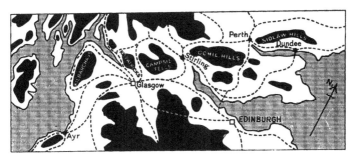

Fig. 41. Sketch-map of Central Lowlands illustrating the control of railway-routes by the relief of the land. The black areas represent land more than 500 feet above sea-level.

by a North British line that crosses the Firth of Forth by the great Forth Bridge, runs through Fife and reaches Dundee by the Tay Bridge across the Firth of Tay. Still keeping along the east coast the line continues to Aberdeen. If a railway were made in a straight line between Glasgow and Perth it would have to rise 1500 feet in the Campsie Fells and over 2000 feet in order to surmount the Ochil Hills. The line between Glasgow and Stirling therefore swings well to the east in order to avoid the Campsies, and then between Stirling and Perth the railway is forced a little west of north in order to circumvent the Ochils.

Glasgow is separated from the fertile Ayrshire plain by an anticline capped by volcanic rocks forming the boundary between Renfrew and Ayr. Fortunately, however, the hills are cut through by two deep valleys which offer easy routes to roads and railways. Through the more easterly of the two (called the **Barrhead gap** from the town at its entrance) runs the main Glasgow and South-Western Railway to Kilmarnock, and thence to Dumfries and Carlisle by the route across the Southern Uplands already described. Another main line of the same railway goes through the western valley to Ayr and Stranraer, with a branch to Ardrossan and Largs. This branch is one of the clearest examples in this country of the control of a route by the relief of the district. From Paisley to Largs the distance as the crow flies is 17 miles. The distance by rail is more than twice that figure, namely 36 miles. The Kilbarchan Hills interpose a natural barrier which it is much easier to circumvent than to surmount. The line, therefore, is forced round the southern extremity of the hills, and then runs due north to Largs.

CHAPTER XVI

THE COUNTIES OF SCOTLAND

Let us consider next the political divisions of Scotland, and see if geographical considerations have had anything to do with the determination of political boundaries. The boundary between England and Scotland is formed by physical features that are barriers to communication. The Solway, the Cheviot Hills, and the River Tweed were important aids to Scotland in her struggles to preserve her

THE COUNTIES OF SCOTLAND

independence. The modern county is a political unit. It is the division of a kingdom administered by a sheriff. The counties were originally governed by the great earls of the country who held the office of sheriff as a hereditary right. The evolution of the county boundaries has been a complex process. They are the final results of a long series of adjustments between forces that were pulling in different directions. The king, the nobles, the church, and the burghs were centres of segregation that frequently did not work in harmony. Yet dominating all these discordant forces the geographical factor is visible. The physical features of the country have moulded the political divisions in harmony with natural regions. In order to make this clear we shall consider the county boundaries in some detail.

Beginning in the south-west of Scotland we find that the southern boundary of **Wigtown, Kirkcudbright**, and **Dumfries** is formed by the natural barrier of the Solway Firth. The northern boundary of all three counties is formed by the highest part of the Southern Uplands. The boundaries of **Ayrshire** are almost entirely formed by hill barriers. Along the south-east the county touches Wigtown, Kirkcudbright, and Dumfries in the highest hills of the Southern Uplands. The boundary that separates Ayr from Lanark and Renfrew is the lava-capped anticline already mentioned as dividing the coalfields of Lanark and Ayr. The three counties of **Lanark, Renfrew**, and **Dumbarton** comprise one natural region, namely, the Clyde Basin. Lanark consists of upper and middle Clydesdale, the southern boundary being a rim of high hills dividing the Clyde drainage from that flowing to the Solway and the Tweed. To the north-west the county ends a little below Glasgow, just where the river becomes too broad to be bridged. Below this point the Clyde is a formidable barrier, and therefore Renfrew occupies one bank and Dumbarton another. Above this point the river can be easily crossed, and therefore both banks are occupied by one county.

In the south-east of Scotland the chief natural region is the basin of the Tweed, and this is occupied by the four counties of **Berwick**, **Peebles**, **Selkirk**, and **Roxburgh**. Another natural unit is the belt of low ground that borders the Firth of Forth on the south. This forms the **Lothians**, which comprise West Lothian (Linlithgow), Midlothian (Edinburgh), and East Lothian (Haddington). In Saxon and in Celtic times these three counties formed one district. The peninsula of **Fife** forms a single unit with well-marked boundaries, the Tay, the North Sea, the Forth, and the Ochil Hills. This county is often called the "Kingdom of Fife," and obviously the name has a geographical significance although given primarily for historical reasons. On the other hand the geographical factor has had little to do with the tiny counties of **Clackmannan** and **Kinross**, the origins of which must be traced in history. Strathmore and the south-eastern slope of the Grampian Highlands are occupied by four counties, **Stirling**, **Perth**, **Forfar**, and **Kincardine**.

In the Highland counties the control of political boundaries by geographical factors is not so clear. The wedge-shaped area between the Dee and Spey is now occupied by the counties of **Aberdeen** and **Banff**, which represent the ancient earldoms of Moray and Buchan. **Nairn** and **Elgin** both comprise a lowland plain bordering the Moray Firth and a hinterland of mountainous country. Glenmore nowhere forms a very formidable barrier to communication, and so we find that it runs through the heart of **Inverness**. **Caithness** and **Sutherland** were formerly so far removed from centres of political control that they were for a long time governed by the King of Norway. This fact is commemorated by the name

THE COUNTIES OF SCOTLAND

Sutherland — southern land. The boundary between Caithness and Sutherland approximates to the geological division between the Old Red Sandstone and the Highland schists. **Argyll** is a historical, not a geographical unit. It is the country of Clan Campbell, and the fact that it encroaches on the territories of other clans is a tribute to the ability of the predecessors of the Dukes of Argyll. **Bute** is the only purely insular county in Scotland. It consists of the islands of Arran, Bute, and the two Cumbraes, and some small islets. All the other islands of Scotland are attached to mainland counties. Why should the islands of Bute not occupy the same subsidiary position? The explanation is found in the early importance of Rothesay, the present county town. It was made a royal burgh in 1400, and its castle dates from the eleventh century. Only a narrow strait separates Bute from Argyll, but it would obviously have been absurd to make Rothesay subsidiary to Inverary, the much less important capital of Argyll. Bute, therefore, has preserved its independence.

CHAPTER XVII

THE BUILD OF ENGLAND

In England as in other parts of the British Isles the fundamental contrast is between the north-west and the south-east. In the north and the west, the old resistant rocks are found that form the highland areas of England, in the south and east are plains or low hills, and the contrast is not one merely of relief, but also of rock-structure, vegetation, industries, and people. The hills of the Cheviots merge without a break into the **Pennine Uplands** which stretch south from the Scottish border to the river Trent. The Pennines were formed by pressure acting in an east and west direction, which ridged the rocks into a great anticline or arch, the axis of which runs north and south. The earth movement took place at the end of the Carboniferous period. We know this because the Pennine anticline consists of Carboniferous rocks, while rocks of younger age are found lying horizontally along both flanks of the arch, showing that they were formed after the earth movements had ceased. Fig. 42 illustrates the arrangement of the rocks. Thick beds of limestone are frequently found in the Pennines, and this gives a very characteristic type of scenery which, however, will be treated more fully in a later chapter when the Pennine Uplands are described in detail.

Examination of an orographical map alone would lead one to believe that the mountainous area known as the **Lake District** was an offshoot of the Pennine Uplands, for the two hilly areas are connected by a neck of high land at Shap. The continuity, however, is accidental. The Lake District is structurally quite distinct from the Pennines, but the two areas have been, as it were, soldered together at Shap by a mass of granite and other igneous rocks, which give them a deceptive appearance of unity. The rocks of the Lake District are much older than those of the Pennines. They are largely of Silurian[1] Age and resemble the rocks of the Southern Uplands and of Wales. Many of the rocks are of igneous origin, and the whole complex mass has been elevated into a dome, and then trenched by rivers and glaciers.

In rock structure **Wales** is very similar to the Lake District. Most of the rocks are of Silurian[1] Age, and

---

[1] Silurian is used with its former wide application, that is, it includes both the Silurian and the Ordovician Systems of modern geologists.

THE BUILD OF ENGLAND 101

consist often of hard grits and slates. The Welsh Highlands form a roughly oblong mass of mountainous country, which in the north attains a greater height than any other part of the British Isles south of the Grampian Highlands.

Originally Wales was a tableland with a moderately uniform surface, but the etching out of valleys has left a tumbled sea of mountain peaks. The evolution of Wales has thus been almost exactly the same as that of the Highlands or Southern Uplands of Scotland. The old rocks that occur in the north and west of Britain are resistant in themselves, but when they are pierced and, as it were, buttressed by igneous rocks an especially obdurate rock complex is formed. Such areas form the highest parts

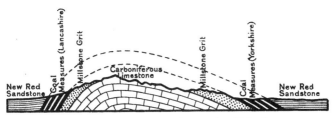

Fig. 42. Section across the Pennine Uplands.

of Britain. The loftiest part of Wales, the Snowdon district, is such an area, so also is the Scafell group, Ben Nevis, and the Cairngorm Mountains. In Ireland, too, the same rule holds.

Old rocks occur again in the south-western peninsula of England. **Dartmoor**, the highest part of this district, reaches in places 2000 feet above the sea. These old rocks are pierced by great bosses of granite that form Dartmoor, the Bodmin Moors, and minor hilly areas of Cornwall. **Exmoor**, which attains almost 2000 feet in height, is composed of hard slates and grits.

The centre of England is a plain, called sometimes the "**Red Plain**" from the colour of the underlying rocks.

The soil is red, the rocks exposed in cuttings are red, and the buildings, being mainly of local stone, are also red. This rock series is known as the New Red Sandstone, and consists of soft red sandstones and friable marls and clays. These materials are easily eroded, and so the district is a plain. The New Red Sandstone series is economically of considerable importance, because practically all our supplies of salt come from these rocks. From the Central Plain an arm of New Red rocks stretches north almost to the Humber, another reaches the Bristol Channel, but the most important of all is the broad band that stretches north-west through Cheshire to the sea. This broad and low passage between the Pennines and the Welsh Mountains offers an easy route from the centre and south of England to the west coast. From the earliest times it has formed one of the most important highways of communication in England. The passage is known as the **Midland Gate**. In a few places the plain of New Red Sandstone is broken by abrupt hills that rise like islands above the general level of the low ground, and present outlines that are more rugged than those associated with the young, soft rocks. These hills are composed of old rocks that are more resistant to denudation. **Charnwood Forest** reaches 850 feet above sea-level. The rocks of these hills are among the very oldest in England. Some of them are of a granite nature and are largely used as road metal. The **Malvern Hills** which rise abruptly from the Red Plain are of Silurian Age.

East and south of the "Red Plain" are the "**scarped lands**" of England. The name comes from the escarpments or scarps that are the most striking features of the scenery. These are long ridges which have a very steep slope or escarpment on one side and gentle dip-slopes on the other. They are caused by the outcrop of a hard band of rock that is not quite horizontal, but has a gentle dip in one

THE BUILD OF ENGLAND 103

direction. Fig. 43 shows how this produces asymmetrical slopes. In England there are two strata in particular that tend to produce escarpments and dip-slopes. One is a band of **hard limestone** which occurs in the Oolitic Series of the geologists, the other is a band of **hard chalk** that forms the uppermost member of the Cretaceous System. Apart from the limestone and the chalk the rocks are mainly clays that are easily worn away, and thus form the low ground between the escarpments.

The **Oolitic Limestone** forms practically a continuous ridge that stretches from the source of the Thames to the north of Yorkshire. The ridge is highest (over 1000 feet)

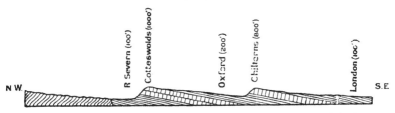

Fig. 43. Section from Wales to London illustrating the formation of escarpments and dip-slopes by bands of hard rock.

at its south-western end, where it is known as the **Cotteswold Hills**. The ridge runs through Northamptonshire, is cut through by the Witham at Lincoln, and in the north of Yorkshire turns east and is cut off by the sea. This part is called the **Cleveland Hills**, and is the richest iron-ore district in England. All along the outcrop of Oolitic Limestone, however, there are valuable deposits of iron ore. The limestone itself is much used for building stone.

The **chalk ridge** runs roughly in the same direction as the one described, and, like the limestone escarpment, has a steep slope to the north-west and a very gentle slope to the south-east. The southern part of the ridge is highest and is called the **Chiltern Hills**. The chalk band can be traced north till it is broken by the Wash. North of the Wash the ridge can be followed again to the Humber. This part is called the **Lincoln Wolds**. The hills reappear again north of the Humber where they are called the **Yorkshire Wolds**. The chalk band then turns east and runs out to sea, making a prominent feature at Flamborough Head.

The chalk stratum dips below London and the lower Thames but reappears in the **North Downs**. London therefore lies in a trough which is called the **London Basin**. The North Downs represent not only the southern rim of the London Basin, but also the northern part of a great arch of chalk that once stretched without a break till it dipped down again in the **South Downs**. There was, as it were, a complete wave of chalk, the axis of which ran east and west. The trough of the wave is the London Basin, the crest was the arch that formerly joined the North Downs and the South Downs. The crest of the arch, however, has been removed by denudation, and the underlying rocks of the Weald exposed. Fig. 44 represents a section from the Thames to Beachy Head, which shows more clearly than any explanation why the steep slope of the North Downs faces south, and why the steep slope of the South Downs faces north. Many of the wells of London are sunk through the London Clay until they reach the chalk where water is generally found. The rain falling on the chalk rim of the London Basin saturates the chalk below London, for the chalk rests on clay which does not allow the water to escape.

The **Weald** was formerly densely wooded; indeed the word *weald* is practically the same as *wold* and the German *wald*, and means a *forest*. The very names of the Weald villages show this. A glance at a large scale map reveals that many of the names end in *hurst*, a wooded hill, or *den*, a forest. The timber of the Weald was largely

THE BUILD OF ENGLAND

cut down for smelting iron in the days when coal had not yet been used for that purpose. On the introduction of coal for smelting, the industry migrated north to the coalfields.

Towards the west (in Hants and Wilts) the London Basin and the Weald Anticline gradually die out, and the chalk stratum there presents a continuous sheet nearly forty miles wide. It is this extensive outcrop of chalk that forms the plateaus and high plains of Hampshire and Wiltshire. Towards the south (in Hants and Dorset) the

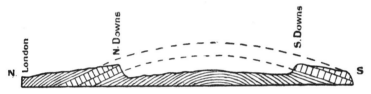

Fig. 44. Section from London to Beachy Head illustrating the origin of the North Downs and the South Downs.

chalk dips again to form another trough known as the **Hampshire Basin**, the southern rim of which comes to the surface in the Isle of Wight and the Isle of Purbeck. The clays of the Hampshire Basin are the same as the London Clay.

CHAPTER XVIII

THE RIVERS OF ENGLAND

IN dealing with the rivers of England we shall spend no time in giving names and positions and lengths of rivers. This information is best obtained from the map. We shall refer to particular rivers merely as a means to illustrate certain **principles of river development.** The way in which certain rivers affect the industries of this country will be mentioned when different natural regions of England are discussed in detail in future chapters. The chief watersheds of England coincide roughly with the most important hills. The Pennines, Central Wales, and the Weald form the main divides. The Oolitic Ridge and the Chalk Ridge are minor divides which, however, are crossed by large streams. As a rule the highest part of the hills is *not* the watershed. For example, the rivers of the south-east of England rise in the Weald and make breaches in the high chalk ridges that intercept them from the sea. A marsh is a commoner watershed than a mountain peak.

The rocks of the English Plain north of the Thames dip generally towards the east or south-east; in other words, the strike or graining of the rocks runs north and south, or north-east and south-west. Rivers flowing east or south-east are, therefore, generally consequent rivers, and rivers flowing at right angles are subsequent. It is frequently found that consequent rivers, which are at first obviously the main streams, are preyed on by subsequent rivers, which behead the consequent rivers and grow at their expense. This is explained by the fact that subsequent rivers run parallel to the outcrops of rock, and a subsequent stream which develops along a particularly soft stratum will grow in a correspondingly rapid manner. We shall find several examples of this principle in the rivers of England.

The tributaries of the **Yorkshire Ouse** (Swale, Ure, Nidd, Wharfe, Aire) are consequent, and may in former times have reached the sea as separate rivers, but the main trunk stream of the Ouse is subsequent, and seems to have cut them off from their lower reaches, and diverted them all into the Humber. Most of these consequent

THE RIVERS OF ENGLAND 107

tributaries of the Ouse formerly extended farther west. The **Aire**, for example, even yet cuts a valley right across the Pennines, but its head-waters seem to have been captured by the **Ribble**. Of all the tributaries of the Ouse, the **Derwent** is perhaps the most peculiar. It rises in the moors south of Whitby and flows south-east to the low Vale of Pickering, whence the route to the North Sea is easy. When only five miles from the sea, however, it turns inland and flows south-west until it joins the Ouse fifty miles away. The explanation of this anomalous course is of fairly wide application, and throws light on several English rivers. The Derwent originally flowed south-east to the sea. In glacial times, however, this outlet was blocked by the great ice-sheet that invaded England from the North Sea and dammed back the Derwent, thus forming a lake. The waters of the lake reached a level high enough to escape towards the south-west over the rim of high ground that bounds the Vale of Pickering in that direction. The escaping waters formed a narrow gorge, through which the Derwent continued to flow even after the retreat of the ice. The theory is supported by evidences of a former lake in the Vale of Pickering. Where anomalous river courses are associated with gorges and lake-like depressions (not an uncommon circumstance in this country) a similar history may be suspected.

In addition to the Ouse the Humber receives the drainage of the **Trent**, the head-streams of which rise to the south-west of the Pennines. In its lower course the Trent flows parallel to the Oolitic Ridge, and in this part of its course is plainly a subsequent river. It is possible, however, that the Trent originally passed through the deep gap in the limestone ridge now occupied by the **Witham** at Lincoln. A subsequent stream from the Humber rapidly pushed back its source and deepened its valley, for it was working along the soft clay outcrops west of the Oolitic Limestone. In time it captured the Trent and diverted most of its waters into the Humber. On this hypothesis the Witham represents the mutilated lower portion of the Trent. The hypothesis is supported by the fact that even at the present time exceptional flooding might inundate Lincoln with Trent water.

One of the most peculiar features of the Chiltern Hills is the number of **gaps** in the ridge, seemingly cut by rivers, but now dry. These gaps are important, for they offer easy routes for railway lines between London and the Midlands. Although the gaps are not now occupied by rivers, as a rule a river begins a little to the south-east of each gap. It is fairly obvious that we are dealing with a whole series of **beheaded consequent streams.** The gaps in the chalk ridge were cut by rivers that originally rose much farther to the north-west. But the development of subsequent streams along the clay vale north-west of the Chilterns has diverted the consequent head-streams into other channels. The chief of these subsequent rivers are the **Thame** flowing south-west and the **Great Ouse** flowing north-east.

The **Thames** presents many interesting and difficult problems. The present river is a very degenerate representative of the former stream, which was much longer and broader. It probably took its rise originally in Wales. The upper Severn and the present Thames may once have formed one river. Indeed in the Thames valley deposits of gravel are found that must have come from the Severn basin, and they do not seem to have been carried by ice. It is the old story again, of late, subsequent streams beheading the original consequent. The lower part of the Severn and the Warwickshire Avon have been the pirates on this occasion. The Thames cuts a very deep and narrow **gap** through the Chilterns at **Goring**. Obviously the river must be anterior to the ridge, or it could not have surmounted it and notched it. The whole surface of the country must have been many hundred feet higher. When the chalk ridge began very slowly to stand out in relief owing to the lowering of the surrounding clays, the river was able to saw down the notch in the ridge as fast as the general level of the clay plains was lowered. An alternative explanation has been offered recently that has not been received without criticism. It is that an ice-dammed lake formerly existed about Oxford, and the overflow

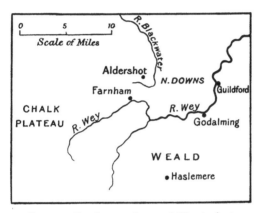

Fig. 45. Sketch-map of part of Wey basin to illustrate river-capture.

drained across the Chilterns and cut the gap at Goring. It will be remembered that a similar explanation was given of the anomalous course of the Derwent in the Vale of Pickering.

The **rivers of the Weald** seem to have originated when the chalk arch between the North Downs and the South Downs was unbroken. It would be difficult otherwise to explain the fact that the rivers rise on the comparatively low surface of the Weald, and cut their way through steep chalk barriers several hundreds of feet high. As the crest of the anticline was lowered by denudation, and the chalk escarpments became higher and farther apart, the rivers kept pace by cutting down the gaps. There is one very good **example of river capture** in the Weald drainage area. The Blackwater rises just north of the North Downs near Aldershot (see Fig. 45). The valley of the upper Blackwater is practically continuous with and less than two miles distant from the valley of the upper Wey. Just at this point, too, the Wey makes a sudden right-angled bend and becomes a subsequent stream. The Blackwater probably came originally from far within the Weald. But the subsequent Wye thrust its head-waters backwards and beheaded the Blackwater near Farnham.

The rivers of the west coast are shorter than those of the east coast, and do not illustrate so simply the principles of river development. The part played by the Severn and the Ribble in tapping some of the water formerly draining to the east coast has already been mentioned. In the extreme north-west the **Eden** is re-excavating a valley of very ancient date. This was a depression in very early times, for the Eden runs down a long thin strip of New Red Sandstone rocks enclosed by Carboniferous rocks. The New Red Sandstone obviously fills up what was a very ancient depression.

CHAPTER XIX

LONDON

LONDON is by far the greatest city in the world. In the Administrative County of London there are $4\frac{1}{4}$ million people, while in Greater London, the area of continuous streets and houses, there are about 7 millions. There are more people in London than in all Scotland or all Ireland. From east to west or from north to south one can travel for over 15 miles and never leave the crowded streets of the great city.

London has been an important town since the Roman invasion of Britain. The place on which it is built has advantages that no other site in the neighbourhood has. One wonders at first why a **situation** was chosen so far up the Thames. Below London the river winds from side to side in a wide, flat, marshy valley, on the sides of which, at low tide, broad stretches of black mud are seen. The city of London was founded on a hill near the edge of the river at a place where there are no marshes on the northern side. On the southern side, too, the marshes very soon rise into higher ground, and so the Thames is easily crossed at London. Traders in olden times soon found that this was a convenient place to cross the river, and the Romans made a road to this point from the Straits of Dover, and built the first bridge across the Thames at London.

In early times practically every trader from the Continent crossed the English Channel where it is narrowest, and landed at Dover. Then the **main trade route** to the heart of England naturally was directed to London, for, as we have seen, the river was easily crossed at this point. The road soon became a famous route and was known as **Watling Street**. It went right through London and away to the north-west as far as Chester. Traces of this old Roman road can be found to this day. But London has other advantages of position. It is the lowest point at which the Thames can conveniently be bridged. Again, the river is tidal to a point a little above London, and this has been of great importance in developing London as a seaport.

On the slopes of the little hill near London Bridge, then, the great city grew up. If one is going towards London Bridge from the west, the slope is quite noticeable when one comes to Ludgate Hill. From the earliest times the part of London near the bridge has been the trading centre, and to this day it is called simply the **City**, a district of famous banks and great offices. London rapidly became so large and important that a royal palace was built near it. A site was chosen about two miles up the river from London Bridge, where there was a little island among the marshes. A great church was built near the palace; to distinguish it from the cathedral it was called the West Minster. Such was the origin of **Westminster Abbey**. Naturally the houses of the nobles, the Law Courts, and the Houses of Parliament were all built near the palace of the King. The sovereign now lives in Buckingham Palace, and the Law Courts are removed to the Strand, but the Houses of Parliament and all the chief government offices are still in Westminster.

There are so many interesting buildings in London that only a few of the most important can here be mentioned. The origin of **Westminster Abbey** has already been indicated. It was founded in the eighth century, and has been more than once rebuilt since then. Since the time of Harold every English sovereign has been crowned in

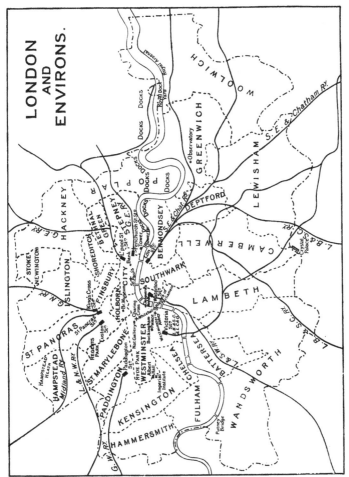

Fig. 46. Map of London.

the Abbey. **St Paul's Cathedral** is the most conspicuous landmark in all London. Its great dome seems to dominate the city. On this spot there has been a church since the beginning of British history. Before there was a Christian church the Romans had a temple to Diana here. The present building, designed by Wren, replaced the cathedral that was burnt to ashes in the Great Fire. Just east of London Bridge is the **Tower of London**, built by the Normans on the ruins of a Roman fort. No other building in Britain could tell so many terrible secrets as this grim fortress. London is, of course, the chief place in Britain for museums, libraries, and picture galleries. There is a great University and many fine colleges.

London is the greatest manufacturing city in the world. Like most large capitals no one industry is pre-eminently associated with it, as metal working is with Birmingham, engineering with Glasgow, the cotton trade with Manchester. Almost everything is made that serves the needs of a great and wealthy community. The factories are naturally found mainly in the east. There is some tendency to segregation of the different industries. Thus **furniture** is made chiefly in the north-east, the enormous **clothing** trade is associated mainly with the east end, the **metal-workers** are in the north-central district, while the **potteries**, the **breweries**, and the **tanneries** congregate on the south side of the Thames. The factories of London have the great commercial advantage of an enormous market at their doors; they are at a disadvantage in their distance from coalfields. Certain industries are therefore leaving London. Shipbuilding is the most noticeable, this industry being now almost extinct on the Thames.

London is by far the greatest seaport in this country. Liverpool, which ranks next, has about three-quarters of the total tonnage entering and leaving London in a year. Food-stuffs are by far the most important of the **imports**.

Fig. 47. The Thames from the Tower Bridge. St Paul's, the Monument, and London Bridge may be seen in the background.

Of the raw materials, wool is three times more valuable than any other import, and among manufactured articles metals are an easy first. London has almost a monopoly of the import trade in certain commodities, although these things may be in common use all over the kingdom. For example, it imports about 98 per cent. of the tea, over 80 per cent. of the rubber, three-quarters of the coffee, and more than half of the wool, skins, tallow, and leather that come into Britain. The explanation of this fact involves many factors, some geographic, some purely economic. One of the main causes is that the commodities mentioned are not (like wood or ores) consigned directly to manufacturers. The trade is in the hands of brokers, who find London the most convenient centre. Another aspect of London's import trade is akin to the foregoing, namely, the fact that London has as large an entrepot trade as all the rest of the ports of Britain taken together. By an entrepot trade is meant the importing of articles not destined to be used by the importing country, but to be re-exported to foreign lands.

Chief Imports and Exports of London, 1911

Imports	Value	Exports	Value
Raw wool	£22,468,000	Cotton goods	£9,134,000
Grain, flour, etc.	16,948,000	Woollen goods	6,408,000
Meat	15,846,000	Iron and steel goods	5,051,000
Tea	12,634,000	Machinery	4,695,000
Sugar	9,250,000	Apparel	2,956,000
Butter	8,937,000	Arms and ammunition	2,151,000
Tin	8,845,000		
Rubber	7,361,000		
Oil	6,523,000		
Timber	5,921,000		
Skins and furs	5,405,000		
Leather	5,152,000		

CHAPTER XX

THE THAMES VALLEY

THE **Thames** rises in the Cotteswold Hills, and flows eastwards across nearly the whole width of England. The upper part of the river from its source to the junction of the Thame below Oxford is often called the Isis. Its most important tributaries are the Cherwell, the Thame, the Colne, and the Lea on the left bank, and the Kennet, the Wye, and the Mole on the right bank. We have already indicated that the Thames was originally longer than it now is, having been probably beheaded by the subsequent lower course of the Severn. The Thames valley itself indicates that it was formerly occupied by a much larger stream. The meanders of a river valley are proportional in size to the volume of water that forms them. In the Thames valley there are big, swinging meanders not now water-covered that show that the present river has shrunk considerably.

The wide plains, through which for the most part the Thames flows, are used for the growing of **crops**, while in the wetter meadows bordering the river **cattle** are kept, for the pasture there is very rich. On the Cotteswolds and Chilterns the soil supports a short, sweet turf which makes these parts eminently suitable for **sheep rearing**. On the chalk slopes of the Chilterns **beech woods** are still found, and formerly were far more numerous. The abundant supply of local wood originated a flourishing **furniture trade**, which is still carried on. **High Wycombe** (Bucks) is the centre of a famous chair-making industry which gives work to over a hundred factories. Chairs from this district may be found in every part of the world. Originally they were made entirely from local beech wood, but now 90 per cent. of the wood used is American birch. Another industry of the Chilterns that has had a similar history is **straw plaiting**, the chief centres of which are **Luton** and **Dunstable**. The straw of the neighbourhood was found very suitable for plaiting, and so the industry began. Local supplies, however, soon proved quite inadequate, and the raw material is now imported largely from China and Japan. Both the industries mentioned are examples of "industrial inertia"; that is, the persistence of an industry in a place that has lost its original advantages. In instances of industrial inertia we nearly always find that purely geographical advantages are replaced by economic advantages. In the case of chair-making and straw-plaiting proximity to an abundant supply of raw material no longer exists. But capital, skilled labour, and developed organisation are strong bonds to keep the industries in their original homes.

Most of the Thames valley is an agricultural region, and in such a district **market towns** grow up at convenient centres for the farmers. These towns have all a certain family resemblance, for they are places where the farmers sell their produce, and buy what is necessary for their farms. Saddlers' shops, ironmongers' shops, and grain stores are conspicuous, and the number of inns is above the average. Almost all the towns of the Thames valley above London are of this nature. Some, of course, have acquired increased importance because they are county towns, or because they have special institutions, or because they have a high degree of nodality, and have become large railway junctions, but the kernel of all is the market town.

The principal town of the upper Thames is **Oxford**, at the junction of the Thames and the Cherwell. Oxford

THE THAMES VALLEY 119

is a market town, but its chief claim to notice lies, of course, in its ancient University, the oldest in the country. The University includes over twenty colleges, most of them very beautiful and interesting old buildings. One of the most important institutions associated with the University is the Bodleian Library, which is one of the world's great libraries, and contains hundreds of thousands of books. The Thames at Oxford is very beautiful, and full advantage is taken of the facilities the river offers for

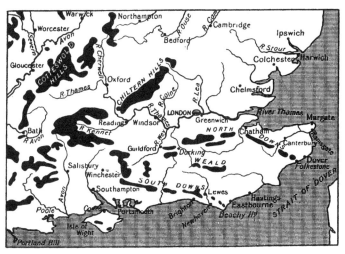

Fig. 48. Sketch-map of South-eastern England. The black areas represent land more than 500 feet above sea-level.

boating. When we remember, too, the numerous historic associations of the town, we must admit that Oxford is one of the most interesting and beautiful towns in Britain. At the junction of the Kennet and the Thames is the town of **Reading.** It is a market town that has become of more importance than most of the others. It has taken advantage of its supplies of grain and flour to become a great biscuit-making town. The nurseries of seedsmen and florists are also conspicuous. Reading has more nodality than any other town in the upper Thames valley, and therefore has become a considerable railway centre. The Thames and the Kennet both offer easy routes through the chalk hills to the west of England, and these are followed by main railway lines diverging at Reading. Another line goes south to the army town of Aldershot.

Why has **Aldershot** become a great military station? The answer is to be found in the character of the ground that stretches north of the town. It is a broad expanse of waste land suitable for exercising troops, but of little value for cultivation. The area is a sandy heath (Bagshot Heath), because the underlying strata are sands of Tertiary age, that is, geologically very recent. Where these sands occur in the London Basin the district is sparsely populated. Other well-known districts similar to Bagshot Heath are Hampstead Heath and Blackheath, identical in origin. These heaths have long been favourite places for recreation, and in recent years have come much into favour as residential sites. Their fresh, open, breezy situation, the healthy soil, and their proximity to London are their chief advantages.

Coming down the Thames from Reading, the first important place we reach is **Windsor**, overlooked by the towers of Windsor Castle. The castle was originally a strong fortress. Since it is within easy reach of London it forms a convenient royal residence. From the battlements of the castle one looks across the river to the playing-fields of **Eton**, the most famous of the great public schools of England. **Harrow**, the great rival of Eton, is also on the Thames valley, and both have the advantage of being comparatively near London. From Windsor to London the Thames is a perfect playground for the people who live in the vicinity of London. On a stretch

THE THAMES VALLEY

of the river just above London the famous boat-race between Oxford and Cambridge takes place every spring.

In the north of the Thames basin the ancient town of **Aylesbury** should be noted. It is the chief centre of a busy dairy-farming district, and sends large quantities of milk to London. Why should Aylesbury and the surrounding district be so much occupied with dairy-farming?

Fig. 49. High Street, Oxford.

Immediately underlying the chalk beds of the Chiltern Hills, and therefore on the north-west flank of the hills, there is a broad outcrop of clay, which stretches from the Thames to the Wash. This clay makes very rich pasture land, and so along this band we find dairy-farming carried on to a large extent. Aylesbury lies in the middle of the clay band, and it has also the advantage of being on one of the main railway lines to London, and therefore it has become an important centre for dairy produce.

Below the Tower Bridge we find great docks dug out of the river banks. The largest ocean steamers, however, generally come up the Thames no farther than Tilbury Docks, more than twenty miles below London. **Greenwich**, on the south bank of the Thames, is now really a part of London, for the buildings are continuous. The associations of Greenwich have always been with the navy and the sea. Its chief institution to-day is the great Observatory. Not only Britain but most other nations now reckon their longitude from the meridian of Greenwich. Adjoining Greenwich is the busy manufacturing town of **Woolwich**. Here is the Royal Arsenal, a great government factory employing many thousand men, who are engaged in making guns, shot and shell, torpedoes, rockets, and similar things. On the estuary of the river are the naval centres of **Chatham** and **Sheerness**, strongly fortified seaports, with royal dockyards that are among the largest and finest in Europe.

CHAPTER XXI

SOUTHERN ENGLAND

In this chapter we shall describe the part of England that lies south of the Thames, with the exception of the south-western peninsula which forms a natural region with a marked individuality. The south-western peninsula is composed of old rocks, and contrasts sharply in many respects with the rest of southern England. The boundary

SOUTHERN ENGLAND

between the two regions is a line drawn from Torquay through Exeter to Bridgwater Bay. In this chapter, therefore, we exclude Cornwall, most of Devon, and the western part of Somerset.

East of the line from Torquay to Bridgwater Bay, southern England is composed of young rocks, which generally are easily worn away and, therefore, form low ground. The one exception is the **outcrop of hard chalk** that is so conspicuous a feature in many southern coast landscapes. The chalk outcrop attains its greatest extent in Hampshire, Wiltshire, and Berkshire, where it forms the high, rolling, breezy downs of those counties. Towards the east the chalk strata at one time formed a great arch, of which only the north and the south limbs are preserved as the **North Downs** and the **South Downs**. West of Guildford the chalk outcrop is very narrow, and so the North Downs contract there to a narrow, sharp, east and west ridge called the Hogs Back. If we draw a line from Salisbury south-west to Dorchester, and another south-east to Arundel, we enclose between these two lines and the coast a triangular area of Tertiary rocks, mainly sands and clays. Along the lines mentioned the chalk dips below this area, which is called the **Hampshire Basin**. The southern rim of the basin is formed by a sharp upfold of the chalk, which outcrops in the Isle of Wight, and determines the shape of that island.

The **outlines of the south coast** of England are conditioned partly by rock structure and partly by tidal action. The North Downs run out to sea in the South Foreland, and form the line of chalk cliffs between Deal and Folkestone. The coast line is salient again where the South Downs reach the sea in the fine promontory of **Beachy Head**. Between these two headlands is a point of quite a different type, namely, Dungeness. It is a low cape of shingle backed by flat marshes. The tidal current running up the English Channel from the west is here suddenly checked by opposing currents, and deposits its loads of shingle in great banks. Selsey Bill is a low shingle point of the same type as Dungeness. The west coast of the Isle of Wight projects in a promontory caused by the narrow chalk band that runs across the island. The **Needles** are pinnacles of chalk that once formed part of the island. Portland Bill is a mass of Oolitic limestone which is joined to the mainland by the Chesil Bank, a tide-borne accumulation of shingle, more than ten miles long. The limestone from Portland makes a good building stone. St Paul's and many other famous buildings are built of this stone.

The **earlier human settlements** in southern England seem to have been on the high chalk downs of Wiltshire and Hampshire. This is remarkable, for at the present time these areas are sparsely populated. The attraction seems to have been the abundant supplies of flints from which primitive man made his tools and weapons. Nowadays the absence of water in these districts is the chief draw-back to settlement. Chalk and limestone are both porous rocks, and water rapidly sinks through them. Even rivers flowing over such districts are apt to disappear underground. The presence of early man on the broad, chalk plateau of southern England is shown by the wonderful monuments found there. **Stonehenge** on Salisbury Plain is the most famous. The ruins consist of a circular group of enormous standing stones situated in the midst of a number of prehistoric burial places. Near Marlborough are the Avebury ruins, even more gigantic, although not so complete, as Stonehenge. On an open space there were two great stone circles and a long avenue. Most of the stones are now levelled to the ground. Close at hand is a huge tumulus of chalk, said to be the greatest burial mound in Europe. The modern towns of this district are all situated in the

SOUTHERN ENGLAND

valleys. The most important are Salisbury on the Avon, and Winchester on the Itchen. **Salisbury** is built about a mile from an old Roman settlement. It is a historic town with a famous cathedral. **Winchester**, as its name shows, dates back to Roman times. More than any other place it is associated with the name of Alfred the Great. For a long time the city was the capital of England. Some parts of its cathedral are considered the finest in Britain.

The Downs of Surrey, Sussex, and Kent, are dry areas, and, therefore, we find few settlements on them. The chalk

Fig. 50. The chalk-cliffs of Southern England. A view of Beachy Head.

beds, however, rest on sands and clays, and so at the foot of the escarpments of the North Downs and the South Downs we find **springs** breaking out. The springs are marked by long lines of villages, many of which are very old. The other controlling factor in the positions of towns in this region, is the **presence of gaps** through the steep, chalk ridges, for these gaps in the ridges afford easy routes for roads and railways to cross the North Downs and the South Downs. Now, of the villages situated on the outcrops of water-bearing strata, that is, at the base of the chalk escarpments, we should expect to find that those situated at the gaps would become most important, and this is exactly what has taken place. The Wey, the Mole, the Darent, the Medway, and the Wye all rise in the Weald, and cut deep valleys through the North Downs, and at the entrance to these gaps stand the towns of Guildford, Dorking, Sevenoaks, Maidstone, and Ashford. Farnham is situated at a gap west of the Hogs Back, through which the London and South-Western Railway passes, while the Brighton line from London Bridge makes use of a gap near Reigate.

Of the "**gap towns**" two may be selected as typical of the others, and described in somewhat more detail, namely, Guildford at the foot of the North Downs, and Lewes at the foot of the South Downs. **Guildford** has been an important place since the time of Alfred, for the routes between London and the south coast were the same then as now. In order to command the route through the gap, a strong castle was built by the Normans, part of which still remains. In modern times the town has become a fairly important railway junction. This change from a strongly fortified place of historic importance to a railway junction is typical of many towns in Britain. The history of **Lewes** is very similar. It possesses an old castle which was one of considerable importance, and near it is the site of the famous battle. A glance at a large scale map will show that three railway lines from the north join near Lewes, and, after passing the South Downs, separate again. Several main roads also make use of the Lewes gap. Lewes is the county town, and it is significant that the county towns of Surrey, Sussex, and Kent are all gap towns.

Canterbury is not a gap town, but it has a nodal site, for it stands where the great historic highway between Dover and London crosses the Stour. The cathedral is the oldest in England, and is probably associated with more famous names than any other outside London. It contains the tombs of Thomas à Becket, Stephen Langton, the Black Prince, Henry IV, and many others hardly less famous.

The towns along the south coast of England are, as a rule, either seaports or watering places. **Margate** and **Ramsgate** have beautiful sandy beaches, backed by fine chalk cliffs. They are situated on a small detached area of chalk. This fact explains why the Isle of Thanet stands out so boldly. **Brighton** and **Eastbourne** are similar, but more fashionable. **Dover, Folkestone,** and **Newhaven** have daily services of fast steamers to Continental ports. Dover connects with Calais, Folkestone with Boulogne, and

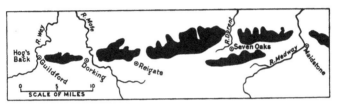

Fig. 51. Sketch-map of North Downs illustrating position of towns with reference to gaps. Black areas represent land more than 500 feet above sea-level.

Newhaven with Dieppe. The distances between these ports should be measured. Dover has been of considerable importance from the earliest times from its proximity to the Continent. In addition, the cliffs at this point are cut through by a valley that gives easy access to the interior. The great naval harbour works recently constructed will add considerably to the importance of Dover.

Southampton is the chief seaport on the south coast of England. It is a town that is growing rapidly, for it has many **advantages for trade**. The busiest commercial sea route in the world is that between the English Channel and New York, and Southampton is obviously very favourably situated with regard to this route. It has the great advantage over its rival on the Mersey, in being near London. Its proximity to the Continent, too, gives it an advantage over all the other great ports of Britain. It has now obtained most of the trade with South America and South Africa. Southampton is also peculiarly fortunate in having **four tides** daily instead of two. One tidal wave comes up the Solent, and another later wave reaches Southampton through Spithead[1]. In the future, Southampton seems likely to grow considerably in importance.

CHIEF IMPORTS AND EXPORTS OF SOUTHAMPTON, 1911

Imports	Value	Exports	Value
Raw wool	£2,574,000	Cotton goods	£5,710,000
Meat	2,521,000	Apparel	2,715,000
Feathers	1,977,000	Woollen goods	1,700,000
Skins and Furs	1,384,000	Leather goods	1,234,000
Butter	1,009,000	Books	582,000
Coffee	632,000		

Portsmouth is the most important **naval station** in Britain. The whole life of the town revolves about the dockyards. Portsmouth Harbour is a well sheltered inlet branching from Spithead. North of the town there is a detached chalk ridge that is very strongly fortified. There are huge storehouses, large dry docks, and several shipbuilding yards. Opposite Southampton Water is the **Isle of Wight**. The island is longest from east to west, because of an outcrop of chalk which forms the upfolded southern rim of the Hampshire Basin. **Cowes** is the most fashionable **yachting** centre in the world. It is not surprising, therefore, to learn that it has a flourishing yacht-building industry.

---

[1] The second tide may also be explained as a reflection from the French coast. This explanation is supported by the fact that a double tide occurs also at Poole.

CHAPTER XXII

THE SOUTH-WESTERN PENINSULA

IN the south-western peninsula we find old rocks, and, therefore, a hilly type of country. The hills occur in bulky masses caused by great intrusions of granite. **Dartmoor** is the largest of the granite bosses, the **Bodmin Moors**, north-east of Bodmin, are composed of another granite mass, while

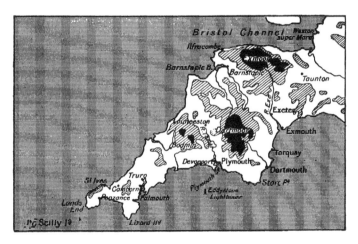

Fig. 52. Sketch-map of the South-western Peninsula. Black areas represent land more than 1000 feet above sea-level; shaded areas represent land between 1000 feet and 500 feet.

smaller bosses form other hills in Cornwall. The extreme tip of the peninsula terminating in Land's End is a granite intrusion. This granite area extended farther west in former times. The **Scilly Isles** are composed of granite and were once connected to the Land's End mass. The hills that terminate in Lizard Point are also composed of igneous rock, although not of granite. The rocks of **Exmoor** are hard sandstones and grits. The Quantock Hills are really a detached portion of Exmoor separated by a narrow valley of New Red Sandstone. The other rocks are slates and grits, all of them older than the Coal Measures.

Of the hilly areas **Dartmoor** is the highest and most extensive. In the north-west of Dartmoor two tors attain a height of over 2000 feet. Most of Dartmoor consists of high, bleak, solitary moors rising up here and there into tors or great blocks of granite, the jointing of which gives the tors the effect of giant masonry. This structure is typical of elevated granite regions. The scenery is stern and impressive, although there are no real mountains as we find in Wales or the Lake District. Dartmoor, like the chalk downs farther east, is noted for its remains of prehistoric man. These are great stone circles, avenues of stones, and giant stone tables. Tools and weapons of stone are also numerous.

The south-western peninsula is rich in **minerals**, which are obtained mainly from the granite areas. The slates surrounding the granite bosses are penetrated by veins containing valuable minerals, of which the chief are **tin** and **copper**. For a long time this district was the richest producer of tin in the world. Two thousand years ago the daring sailors of Tyre and Carthage came for Cornish tin. But many of the mines are now worked out, and the discovery of rich tin deposits in the East Indies has reduced the importance of English tin. The amount of copper produced nowadays is insignificant compared with what we have to import. In places the granite has rotted away until there is nothing left but a thick deposit of soft, white clay, called **kaolin**, which is used in the manufacture of porcelain. The factories of Lancashire also utilise it in the preparation of cotton cloth, so that the bulk of the kaolin is sent by boat to the Mersey which is convenient both for Lancashire

Fig. 53. King Tor.

and the pottery district of North Staffordshire. Fowey in Cornwall and Teignmouth in Devon are the towns that export most kaolin.

The **climate** of the south-west of England is unusually mild, particularly in winter time. Frost and snow are rare, and, therefore, this part of the country has obtained note as a winter resort, especially for invalids. Palms and other southern plants flourish in the open air as they do nowhere else in England. The favourable climate has fostered a considerable industry in growing early flowers and vegetables. The **Scilly Isles** are particularly noted for this industry. As well as being mild the climate is moist, and, therefore, the pasture-lands are exceptionally rich. **Cattle-rearing** and **dairy-farming** are largely carried on both in Cornwall and Devon. In the east of Devonshire, where the old rocks are replaced by the New Red Sandstone, the land is lower and flatter, and the rainfall is less. The fertile soil of this district is very suitable for **fruit-growing**. Devonshire is famous both for cream and for cider. The **fisheries** of the south-western peninsula have been mentioned in a former chapter. Mounts Bay is the centre of the spring mackerel fishing, and Penzance is then at its busiest. The herring boats make St Ives their principal headquarters, while the deep sea trawlers sail from Plymouth. The pilchard fishing was formerly of great importance, but has declined very much in recent years.

The most important towns of the south-western peninsula are situated at the mouths of rivers, where the sea-cliffs are broken by estuaries that form good harbours. Thus Plymouth, Dartmouth, Exmouth, Falmouth, Teignmouth, Fowey, and a hundred others are situated at the mouths of rivers. A glance at the map shows that these estuaries are long, branching channels, running far into the land and forming splendid natural harbours. The peculiar shape of these estuaries shows that they are simply **drowned river**

THE SOUTH-WESTERN PENINSULA 133

mouths, formed by a subsidence of the land. They gradually widen and deepen seawards, and, therefore, belong to the ria type of drowned valleys rather than to the fiord type. The absence of hollowed-out basins, such as are found in the Scottish lochs, is probably due to the fact that the south-western peninsula, unlike the other highland areas of Britain, was not glaciated during the great Ice Age.

The triple town of **Plymouth**, **Devonport**, and **Stonehouse** is situated on the splendid harbour of Plymouth Sound. It is a naval station and a port of call at which ocean steamers land their passengers and mails for London. From the ridge of Plymouth Hoe a fine view is obtained of the Eddystone Lighthouse. Plymouth labours under a disadvantage that is shared by all the seaports of the south-western peninsula. It is too far from the great centres of industry, in other words it lacks an important hinterland. **Exeter** possesses a considerable amount of nodality. We might have guessed this from the fact that it was an important Roman station. Railway lines radiate in all directions. The town possesses a fine cathedral. **Torquay** on Tor Bay is the most popular of the coast watering places. Its wonderfully mild climate makes it very suitable for invalids in winter. **Ilfracombe**, on the north coast of Devonshire, is another favourite holiday resort. Many of the fishing villages round the coast have discovered that catering for visitors pays better than fishing, and are developing rapidly as watering places. **Penzance** and **St Ives** are among the most popular. **Truro** and **Camborne** are the principal mining centres.

Devonshire has long been noted for the number of **famous seamen** who have been born there. No other county in the kingdom can boast of so many illustrious names. Drake, the greatest of the Elizabethan seamen, was a Devonshire man, so was the courtier, poet, soldier, and sailor, Sir Walter Raleigh. The valiant Hawkins, slave trader and hero, was born in Plymouth. The terrible Grenville, as fierce as he was brave, was another son of Devon. Sir Humphrey Gilbert, Raleigh's half brother, and John Davis of Arctic fame, were born near Dartmouth. One cannot help thinking that geographical conditions have been a factor in the fostering of so many famous seamen. Devonshire is part of a peninsula with a long seacoast with good harbours, but more than this is needed to produce sailors. If the country is rich and flat, there is no inducement for its inhabitants to leave it, but if it is sterile or hilly, the people are forced on to the sea and become a race of sailors. Greece, Portugal, and Norway splendidly exemplify this tendency, and on a smaller but no less marked scale Devonshire also illustrates the principle.

CHAPTER XXIII

THE SEVERN VALLEY AND THE BRISTOL CHANNEL

THE **Severn** rises on the slopes of Plynlimmon, and for some distance flows north-east. It seems to be making for the Irish Sea. In this part of its course the Severn has all the characteristics of a mountain stream. Its flow is rapid and its valley is narrow. Above Shrewsbury it turns east, and below Shrewsbury its course alters to south-east, and then almost to due south. The change in direction is accompanied by a change in the character of the valley. Near Shrewsbury the valley is wide and open, and the river flows in meanders like a stream nearing the sea. Below Shrewsbury the valley contracts again, until at Ironbridge it has become a gorge with the hills of the Wrekin on the left bank and the long ridge of Wenlock Edge on the right

THE SEVERN VALLEY

bank. The Severn has become an upland river again, and keeps this character until above Worcester it finally becomes a lowland stream.

The changes in the nature and direction of the Severn present a number of difficult problems to the physical geographer. In a former chapter we saw that the upper Severn may formerly have formed part of the Thames system until captured by the subsequent river now represented by the lower Severn and the Avon. It has been suggested that the part of the Severn above Shrewsbury originally flowed to the Irish Sea, but during the Ice Age this course was dammed, and a lake formed near Shrewsbury. The waters of the lake escaped over the rim of hills to the south, and cut the gorge at Ironbridge. On the retreat of the ice this course persisted. In any case we have in the upper Severn the curious association of gorge and lake-like depression such as we have already mentioned at Pickering and in the case of the Thames.

The upper Severn flows over the old, hard rocks of Wales. When the river turns east above Shrewsbury it enters the plain of New Red Sandstone that forms so much of the Midlands of England. Henceforth to the Bristol Channel the Severn flows over this rock series with very few interruptions. These rocks are of importance economically, for they contain **beds of salt** which furnish nearly all the salt produced in this country. A deep hole is bored and a double pipe put down the bore. One pipe is inside the other and much smaller so as to leave a space round the small pipe. Fresh water is then made to flow down the outside pipe, until it reaches and dissolves the salt deposits. The brine is then pumped up the inner tube, and the salt recovered by evaporation. **Droitwich** is the chief centre of the salt industry.

The mouth of the Severn is shaped like a long funnel. As the tide comes up this funnel-shaped estuary the water is hemmed in closer and closer and tends to become heaped up, until at last a swiftly moving wall of water is formed. This tidal wave is called the **bore** of the Severn, and often reaches five or six feet in height, so that small boats are sometimes overturned. The bore is a hindrance to the navigation of the lower part of the Severn, but this has been obviated by the construction of a ship-canal at Gloucester.

In the upper basin of the Severn **Shrewsbury** is the most important town. It is a picturesque old town with steep, narrow streets and timbered houses. Like most of the towns on the Welsh border Shrewsbury has seen much fighting. Its old castle dates from the time of the Normans. Shrewsbury is an excellent example of that change so common in this country, namely, the growth of an old, historic, fortified town into an important railway centre. Railways radiate out in all directions from the town. In its middle course the Severn is joined by the Stour from the Midlands. **Stourbridge**, on the edge of the Black Country, is near valuable deposits of fire-clay. **Kidderminster** is a few miles from the junction of the Stour and the Severn. Carpet-making was started here in the first half of the 18th century. When steam was applied to machinery the town was luckily near enough to a coalfield to be able to develop unhampered by lack of fuel. Lower down the Severn is **Worcester**, an old historic town. As its name indicates, the place was probably a Roman settlement. It is a cathedral city, and industrially is known mainly by its fine china.

At the mouth of the Severn is the seaport town of **Gloucester**. Difficulties of navigation have prevented it from becoming a great seaport, although it is the nearest port to a densely populated part of the Midlands. The trade of Gloucester is curiously one-sided, being almost entirely an import trade. The value of the imports is often

AND THE BRISTOL CHANNEL

more than a hundred times that of the exports. The chief imports are grain and timber. Iron goods, coal, and salt are exported. Gloucester is the nearest port to Birmingham and the Black Country, but comparatively little of the goods of that district are exported from Gloucester. The exports of the Midlands are costly manufactured articles which are sent to ports with better facilities than Gloucester. The extra cost of carriage is small compared with the total

Fig. 54. The Bore ascending the Parrett from the Bristol Channel.

cost of the articles. On the other hand grain and timber are bulky and heavy, and are sent by the cheapest route.

CHIEF IMPORTS AND EXPORTS OF GLOUCESTER, 1911

Imports	Value	Exports	Value
Grain	£1,952,000	Salt	£6,000
Timber	548,000	Coal	4,000
Sugar	527,000	Iron and Steel goods	2,000
Total Imports	£3,523,000	Total Exports	£24,000

The county of **Hereford**, on the borders of Wales, is

drained by the Wye, which flows into the mouth of the Severn. This district presents a striking contrast to its surroundings, on the north-west the barren uplands of Wales, on the east, south, and south-west grimy, industrial regions. The Wye valley is sheltered from rain and wind by the mountains of Wales, and the climate is therefore unusually dry, so that the **orchards** and **hop-gardens** rival even those of Kent. The fruit-growing industry of this region is a good illustration of the value of rapid communication. Before the district was opened up by railways the industry was of merely local importance. Nowadays special fruit trains are sent to all parts of Britain. Every week in the season thousands of tons of fruit are despatched in this way.

Just as Liverpool is the natural termination of the route from London to the west coast through the Midland Gate, so **Bristol** is the termination of another easy route. The chief obstacle between the Thames valley and the Bristol Channel is the high limestone ridge of the Cotteswolds, and this is deeply cut through by the Bristol Avon. The gap forms a channel for trade from the south-east of England to the Irish Sea and the Atlantic Ocean. A band of Carboniferous Limestone outcrops just at Bristol, and the Avon cuts its way through the hard rock in the deep Clifton Gorge. On both sides of the river the cliffs rise in sheer limestone walls two hundred feet high. Bristol is one of the oldest seaports in the country. Here John Cabot put to sea in command of the expedition that first discovered the mainland of North America. In those days Bristol had a large trade in slaves. Its position favoured the growth of **trade with Ireland and America.** Indeed it was the first seaport in the country to secure a large Atlantic trade, and for many years it was the biggest port in Britain, after London. The trade of Bristol is still carried on chiefly with Ireland and America. **Tobacco** and raw **cocoa** are imported, and there are large factories for manufacturing these articles. From the West Indies and Central America come sugar, bananas, and pineapples. All the bananas entering this country come either to Bristol or Manchester, and are thence distributed. The huge ships of to-day are not able to come up the Avon to Bristol, and so great docks have been built and an "out-port" formed at Avonmouth. The factories of the district obtain their fuel from the Bristol Coalfield which stretches north and south just east of the town. Bristol is connected with Wales by the **Severn Tunnel**, the longest in Britain. The town, therefore, lies in the direct route between South Wales and London.

CHIEF IMPORTS AND EXPORTS OF BRISTOL, 1911

Imports	Value	Exports	Value
Grain	£4,906,000	Iron and Steel goods...	£2,414,000
Sugar	1,266,000		
Cheese	1,006,000		
Fruit	936,000		

Higher up the Avon is the town of **Bath**, which has been a favourite health resort from the time of the Romans, who built baths to utilise the hot springs of the district. The situation of Bath resembles that of, say, Guildford, for it lies at the foot of a steep escarpment and at the mouth of a gap cut by a river. The mild climate, the health-giving springs, and the picturesque situation of Bath make it one of the most popular inland watering places in England. Ten miles south of Bristol an outcrop of Carboniferous Limestone forms the **Mendip Hills**. They are not separate hills, but a plateau with steep sides and a level surface. Perhaps the most striking features of the Mendips are the wonderful caves and gorges near Cheddar. These are characteristic of all upland limestone regions, but the Cheddar gorges are the finest in England. The cliffs rise in perpendicular walls for over 400 feet. South of the

Mendips the New Red Sandstone plain of Somerset is a noted pastoral district. Its Cheddar cheese is famous.

The West of England **woollen industry** deserves some mention. Raw material was obtained from the sheep reared on the grassy pastures of the Mendips, the Oolitic Limestone Ridge, and the Chalk Ridge. In addition there is in the neighbourhood a thick bed of fuller's earth, much used for cleaning wool. A flourishing cloth-making industry sprang up at Bradford, Trowbridge, Frome, and Stroud. The last named town brings its coal across the Severn from the Forest of Dean Coalfield, while the other towns are nearer the Bristol Coalfield.

CHAPTER XXIV

THE MIDLANDS

MOST of the centre of England is occupied by a plain which is known as **the Midlands**. The term is often used in rather an indefinite way, but there are certain definite physical features that we may take as forming the natural boundaries of the Midland Plain. On the west the Midlands stretch to the Severn, while on the south and east the Oolitic Limestone Ridge forms a sharp boundary. At Cheltenham the limestone ridge curves in an easterly direction and passes just west of Northampton. Thence it runs almost due north to Grantham and Lincoln. The Trent and the southern hills of the Pennines may be taken as the northern limit. On the north-west the plain extends without interruption to the Irish Sea. This broad, low tract, stretching to the estuaries of the Dee and the Mersey, is called the **Midland Gate**. The principal rivers

THE MIDLANDS

of the Midlands are the Warwickshire Avon and the Trent with its tributaries.

The **rocks** of the Midlands belong mainly to the New Red Sandstone system. These rocks weather easily, and

Fig. 55. Typical limestone scenery. The Cheddar Gorge.

thus give rise to plains where they occur in broad areas, or to valleys where they are found in narrow strips between harder rocks. In a former chapter we noticed that the red plains of Somerset and Devon furnished rich pastures on which numerous cattle were kept. This is true also of the Midlands. Where the New Red rocks occur, the number of cattle reared is above the average. For example, Cheshire, in the extreme north-west of the Midland Plain, has more cattle for its size than any other county in England. **Dairy-farming** is one of the chief occupations in this shire, its cheese being especially famous.

Although the Midland Plain is mainly occupied by the New Red Sandstone, this is not the most important rock series. There are several areas where the Coal Measures outcrop, and these rocks are of course economically the most valuable. There are two fairly large and important coalfields and two smaller ones. The large fields are in North Staffordshire and South Staffordshire respectively. One of the small coalfields is called the Warwickshire Coalfield and stretches from Tamworth to Nuneaton. The other surrounds Ashby-de-la-Zouch in Leicestershire.

The **North Staffordshire Coalfield** contains thick beds of clay, that can be used for making earthenware, and, therefore, the towns of this district manufacture enormous quantities of **pottery**. There is more earthenware made here than in any other area of equal size in the world, and so the whole district is known as the Potteries. Fine clay is imported from the granite areas of south-west England. The chief towns of this district—Stoke, Hanley, and Burslem—have recently been formed into one municipality under the name of **Stoke**. In Burslem was born the man who practically created this enormous industry, the famous Josiah Wedgwood who lived in the second half of the eighteenth century.

The **South Staffordshire Coalfield** is a densely populated area which contains a few large cities and a very large number of industrial towns and villages. On all sides can be seen coal-pits, iron-mines, blast-furnaces,

THE MIDLANDS

forges, and rolling-mills, which pour forth dense volumes of black smoke from innumerable chimneys, and earn for the district its name of the **Black Country**. Along with the beds of coal there are found valuable deposits of black-band ironstone, and therefore this area has become the greatest **iron-working district** in Britain. Pig-iron, steel, nails and chains, screws, guns, pins and needles, tools, motor-cars—these are but a few of the countless things made in the Black Country. **Wolverhampton, Walsall,** and **Dudley** are the names of some of the many large towns engaged in iron-working. In addition, there are large numbers of smaller villages, all engaged either in coal-mining or in some branch of the hardware industry.

Birmingham, the "Capital of the Midlands," is the third largest city in Britain and requires special mention. It lies a little way to the east of the South Staffordshire Coalfield, for it grew up when wood fuel was used for smelting iron. In addition, it is a long way from any important seaport. These two circumstances make Birmingham unable to compete with other towns in crude processes such as iron-smelting, where large quantities are handled at a small cost. The **value of the workmanship** put into Birmingham goods is the chief factor in the cost of the finished article. In other words, compared with weight, the value of the goods is great, and so railway rates are not important compared with the total cost of making the goods. We have already had an illustration of this important geographical principle. In dealing with the Clyde valley we saw that Paisley and other Renfrewshire towns were some way from the Lanarkshire Coalfield, and so specialised in costly machine tools. All kinds of metal goods, from a screw-nail to a locomotive, are made in Birmingham. Sometimes Birmingham **fancy goods** are sneered at as "Brummagem ware," but one should think rather of the amazing skill and organisation that make these goods so wonderfully cheap. Some of the finest jewellery made in Europe comes from Birmingham. The city possesses a modern and well-equipped university. Its water is brought from the heart of Wales through a pipe line a hundred miles long. It comes from two artificial lakes in the upper basin of the Wye.

Coventry on a tributary of the Warwickshire Avon resembles Birmingham in being some distance from a coalfield. We expect therefore to find a trade in highly manufactured articles. It was (and still is) a great cycle-making centre, and so naturally developed into the most important motor-car building town in Britain. Other two towns on the Avon are interesting, not because of their industries, but on account of their history. **Warwick** has a magnificent old castle associated with many famous names, and primarily with the great Kingmaker. Nowadays, Warwick, like most other county towns, is the marketing centre for the surrounding farming district. Lower down the river, the quaint, old-fashioned town of **Stratford-on-Avon** was the birthplace of Shakespeare.

Although nearness to coal is the chief factor in British industries there are other advantages which have caused certain industries to become settled in certain places. Some industries are largely dependent on the quality of the water-supply. **Burton-on-Trent** has some of the largest breweries in the world. To make good beer, the water should contain certain substances in solution, and the water from the deep wells of the neighbourhood possesses the necessary degree of "hardness." On the New Red Sandstone soils of Northampton, Leicester, and Stafford, large numbers of cattle are reared. As a result of this we find that the towns of **Northampton**[1], **Leicester**,

---

[1] Strictly speaking the town of Northampton is not in the Midlands, for it lies to the east of the limestone escarpment which runs through the middle of Northamptonshire.

THE MIDLANDS

and **Stafford** are all important boot-making centres. On the hills that form the eastern border of the Midlands sheep are reared, and so **Leicester** and **Nottingham** manufacture **hosiery.** The latter town is also an important lace-making centre. It obtains its fuel from the Yorks, Derby, and Notts Coalfield which is not far away. The nearness of coal also explains the fact that in Nottingham we again encounter engineering industries.

Several important railway lines cross the Midland Plain. It is obvious that towns on a main line have a better chance of growing than those not well served by railways. Some of the Midland towns owe most of their importance to railways. **Crewe** is a good example of such a town. Main lines radiate from it in all directions. Its importance has been much increased by the establishment of the chief locomotive-building works of the London and North-Western Railway. **Derby** is similar in many respects. It is on the main Midland line, and contains the locomotive-building works of that railway. **Rugby,** too, has benefited largely from the fact that the London and North-Western Railway passes through it. This line passes the Oolitic Ridge by means of a tunnel, and enters the Midland Plain near Rugby. In recent years the town has developed as an engineering centre. Obviously the disadvantages of Birmingham are even more pronounced in Rugby, and we may therefore expect its products to embody a high degree of workmanship and technical skill. This is found to be the case, for it specialises in electrical machinery.

CHAPTER XXV

EASTERN ENGLAND

UNDER this heading we shall describe that part of England bounded on the south by the Thames valley, on the west by the Oolitic Escarpment, and on the north by the Humber. The east of England is on the whole a flat country. Both the chalk ridge and the limestone ridge run through it, but the hills made by these hard bands are not so high as they are near the Thames. No part of the land reaches six hundred feet above the sea. We can trace the chalk ridge through Suffolk and Norfolk where it is called the **East Anglian Heights**. The ridge-like character, however, here largely disappears, for the dip of the strata is very small, and the chalk outcrop is therefore broad and undulating rather than narrow and scarped. The general tameness of the relief is increased also by the fact that all this district is thickly covered with boulder-clay deposited during the Great Ice Age. The chalk outcrop is broken by the Wash, but north of that opening it can be traced continuously to the Humber. In this part of its course it is called the **Lincoln Wolds**. The limestone ridge runs without a break through Northamptonshire to the town of Lincoln, and thence due north to the Humber. This part of the limestone ridge is known as the **Lincoln Heights**. Between the Wash and the Humber the outcrops of limestone and chalk regain their character of sharp ridges. The dip increases, the outcrops are therefore narrower, and the two ridges approach each other closely.

In a previous chapter we saw that west of the chalk band there is a band of clay forming the Vale of Aylesbury.

EASTERN ENGLAND

In eastern England this band of clay widens out and forms most of the low ground round the Wash. This low-lying, flat country is called the **Fens**. Not many centuries ago this land was under water, but mud and sand are continually being brought from the north by marine currents, and deposited round the shores of the Wash. These southward moving, silt-bearing currents are actually pushing the mouths of the rivers towards the south. The Yare now

Fig. 56. Ruins of Dunwich Church.

reaches the sea two miles south of Yarmouth. Formerly the mouth was three miles north of the town. The **Norfolk Broads** are peculiar features of the coast. The mouths of the rivers are narrow, but immediately behind they widen into broad reaches, which are especially well-developed near Yarmouth. The Broads are the happy hunting-grounds of yachtsmen and fishermen.

In the never-ending struggle between sea and land the coast round the Wash is winning the battle. But not far away the waves are victorious, for the **coast** is being continually **cut back.** For example there is a little fishing village called Dunwich on the Suffolk coast that has had a strange history. It was an important place in Saxon times, and was once the largest town in Suffolk. But the sea crept steadily in and undermined the cliffs on which Dunwich was built. One by one, houses, churches, and monasteries found themselves on the very brink of the cliff, then their foundations were cut away, and they collapsed into the sea. One or two of the old buildings still remain. The photograph on p. 147 shows the ruins of an old church on the very edge of the cliffs. It is doomed to destruction. Sooner or later the waves will undercut the cliff and engulf the church.

The east of England is the most important **agricultural district** in Britain. This is due to a number of causes. The climate is exceptionally dry and sunny, there are no hills of any great height, and the soil is good. As a rule the soil is a fairly stiff clay, which is eminently suitable for grain growing. It is not surprising, therefore, to find that Cambridgeshire has a greater proportion of cultivated land than any other county in Britain. Cambridge, Norfolk, Suffolk, Lincoln, and Essex are all famous **wheat-growing** districts. Norfolk is also particularly noted for its **barley.** Large quantities of **turnips** are also grown. The **fisheries** of the east coast are the most valuable in this country, and practically all the towns along the coast are engaged in fishing. These towns are the nearest centres to the North Sea fishing banks, which are the most valuable in Europe. Many thousand men are engaged in this industry. In the autumn herring season thousands of Scotch fishermen come south and take up quarters in the coast towns.

Since most of eastern England is devoted to farming there are not many large towns. The chief towns can be arranged in three groups, coast towns that are engaged in

EASTERN ENGLAND

fishing or are seaports, market towns that distribute supplies to the surrounding agricultural region, and, thirdly, towns that have become important by being situated on main railway lines. There is one town that does not fall into any of these classes, namely, **Cambridge**, which possesses one of the two great universities of the country. As in Oxford, the life of Cambridge centres mainly round the colleges, and the chief interest of the town lies in the fine, old buildings and their beautiful gardens. Boating on the Cam is a favourite pastime. The sport has been moulded by its environment, for the Cam is narrow, and therefore "bump" races have been evolved. But Cambridge was a fairly important place even before the founding of its university in the twelfth century. Its early importance was due to purely geographic causes. It is situated among the fens where in former days communication was very difficult. But just at Cambridge the high areas on each side of the Cam valley come close together, and so make this spot the most suitable for crossing the fens. Two important Roman roads, therefore, crossed the fens at this point, and were guarded by a military station which was the beginning of the town of Cambridge.

Of the coast towns **Harwich** is an important packet station for the Continent. It has two considerable advantages. It is situated on the best natural harbour between the Thames and the Humber, and it can be reached in an hour and a half from London. Harwich connects with the Hook of Holland, whence express trains run to Berlin, and thence to Moscow and St Petersburg. There is not much export trade at Harwich, but there is a very large import trade, chiefly of perishable articles such as butter, eggs, meat, fish, and poultry, and high grade fancy-goods from the Continent.

Lowestoft and **Yarmouth** resemble one another in several ways. They are the chief centres of the English herring fisheries. The herring boats are "drifters" using surface nets. The nets are fastened together, and are sometimes a mile long. Thousands of fishermen have their headquarters in these two towns. Both towns also are popular holiday resorts. All along the coast, indeed, the little towns are catering more and more for summer visitors. Of all the fishing towns **Grimsby** on the Lincoln coast is the most important. The fish brought into Grimsby are mainly of two kinds, those caught by line and hook, such as cod, whiting, and haddock, and flat fish such as turbot and sole, which are obtained by trawling. Every day train-loads of fish are sent to London and the great towns of Yorkshire and Lancashire. Grimsby, too, has a considerable trade with the Continent, and ranks as the sixth seaport in England.

CHIEF IMPORTS AND EXPORTS OF GRIMSBY, 1911

Imports	Value	Exports	Value
Butter	£3,374,000	Cotton yarn	£4,538,000
Cotton goods	1,629,000	Woollen goods	4,503,000
Woollen goods	1,434,000	Cotton goods	3,692,000
Timber	703,000	Wool	1,299,000
Meat	580,000	Coal	951,000
		Machinery	934,000
		Fish	840,000

The second class of towns we mentioned includes those market towns that form the most convenient centres for the farming districts surrounding them. The largest of these, as we might expect, are also the county towns. In Suffolk **Ipswich** is the chief town, situated where the London-Norwich road crosses the Orwell. Its manufactures are typical of a market-town, and include agricultural implements and artificial manures. **Norwich** is the chief town of Norfolk. It is the greatest grain market in this country. We mentioned before that Norfolk was a famous barley-growing district, and so we find that Norwich

EASTERN ENGLAND 151

contains several breweries. It has also manufactures of mustard and starch.

Lincoln is perhaps the best example of a **market town** in the east of England. As its name indicates, it has been an important place from the time of the Romans. The Oolitic Limestone Escarpment runs north and south through the middle of Lincolnshire, and forms an obstacle for traffic between the east and west of the county. Just at Lincoln, however, the ridge is cut through by the River Witham, and an easy route is formed through the gap.

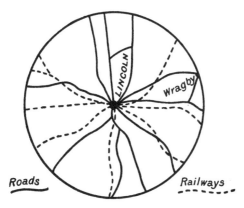

Fig. 57. Routes round Lincoln illustrating the "nodality" of the town.

Several roads unite at this gap, and railway lines radiate from the town in all directions. One may be sure that such a situation would be of great military importance in former times. This opinion is confirmed by the fact that there has been a strong castle at Lincoln for many centuries. Lincoln is obviously the most convenient centre from which to reach all parts of the county, and so it has become the county town. There are large factories, too, for the making of agricultural implements, traction-engines, steam road-rollers, and similar necessities of a farming district. The beautiful cathedral is one of the finest in Britain.

The **names of the villages** of Lincolnshire are interesting, because they show that the Danes once had a very strong hold on this part of England. A large-scale map of the county will show scores of names ending in *by*, which is a Danish termination meaning *a town*. The frequency with which it is found indicates the remarkable number of Danish settlements there must have been. The way in which place-names throw light on conditions that existed long ago is shown also by the towns of the fen country. A large number of names there end in *y*, or *ey*, or *ea*, such as Ely, Welney, Whittlesea. These endings mean *an island*, and they tell us that in former times these places were surrounded by water or by marshes. It will be recalled that the same termination is very common in the Scottish Hebrides, once part of a Scandinavian kingdom.

The last class of important towns that we shall mention consists of those that have grown much in recent years because of their **situation on main railway lines. Bedford** and **Peterborough** are good examples. They are both old towns dating back to Saxon times. They are market-towns, and have always had a trade in farm-produce, but recently factories have sprung up, chiefly for the making of farming implements. Peterborough is now one of the greatest grain-markets in England. The modern growth of these two towns is largely due to the fact that Bedford is on the main Midland Railway line to London, and Peterborough is on the main line of the Great Northern Railway. Both towns, too, have points of interest apart from their trade. Peterborough has a very beautiful cathedral, while Bedford at once recalls the name of John Bunyan.

CHAPTER XXVI

THE LAKE DISTRICT AND THE PENNINE UPLANDS

THE Lake District and the Pennine Uplands are joined at Shap by a neck of high ground 1000 feet above sea-level. Geologically, however, the two areas are quite

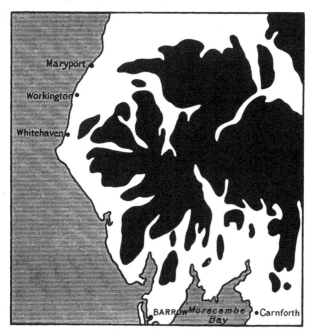

Fig. 58. The Radial Drainage of the Lake District. The black areas represent land more than 500 feet above sea-level.

distinct. The rocks of the **Lake District** are much older than those of the Pennines. They are similar to the rocks of the Southern Uplands and of Wales. They belong to the Silurian system, and consist, broadly speaking, of a northern band of slates and grits, a central band of igneous rocks, including intrusions of granite, and a southern band of slates and grits. These rocks are resistant to the weather, and therefore the Lake District is mountainous. The hilly area is roughly circular in shape, and about 40 miles in diameter. It is believed to have formed originally a great dome which is now deeply etched by valleys. In brief, it has the structure of a "dissected dome."

From the centre of the dome the valleys radiate like the spokes of a wheel (see Fig. 58). It was Wordsworth's keen eye that first perceived this analogy, and this, too, before the days of Ordnance Survey maps, the basis of all modern British maps. Between the valleys the land rises into many fine mountains, chief of which are **Scafell**, **Helvellyn**, and **Skiddaw**, all over 3000 feet high. The peaks and pinnacles about Scafell and Great Gable are well known to mountaineers, who find there very difficult rock-climbing. The influence of rock-structure on scenery is very marked in the mountains of the Lake District. Skiddaw, which is composed of the softer slates, is smooth and rounded in outline, while the harder igneous rocks round Borrowdale and Scafell present a much more rugged type of scenery.

The district gets its name from the number and beauty of the lakes it contains. The lakes are of all sizes, from **Windermere**, twelve miles in length, to the smallest mountain tarn. The large lakes all occur in main valleys. It seems most likely that basins were scooped out of the valleys by the glaciers of the Ice Age, and, on the melting of the ice, these became lake basins. This explains the long, narrow shape of most of the lakes. **Derwentwater** is not particularly narrow, but if we imagine it joined to Bassenthwaite Lake, as it undoubtedly was in former times, the restored lake fully merits the description "long and narrow." Both the latter-named lakes are fairly shallow,

AND THE PENNINE UPLANDS 155

with an average depth of less than 20 feet. **Wastwater**, on the other hand, is very deep. Although the surface of the lake is 200 feet above sea-level, the bottom is nearly 60 feet below the level of the sea. The islands of Derwentwater are composed of loose stones, and seem to be the remnants of moraines. The islands of **Ullswater** are

Fig. 59. View of Derwentwater and the slopes of Skiddaw. Bassenthwaite Lake is seen in the background.

solid rock rising from a great depth, and are smoothed and striated by glaciers. In many places the rapid torrents from the hill sides bring loads of stones and silt, which accumulate in the lakes as deltas. In time these deltas will be pushed right across the lake, and thus cause a division into two smaller lakes. This has already happened in the case of Derwentwater and Bassenthwaite Lake, and also in the case of Buttermere and Crummock Water.

One of the principal charms of the Lake District lies in its beautiful waterfalls. Of these the **Falls of Lodore**, near the head of Derwentwater, so vividly described by Southey, are the best known. The Lodore Falls, like most of the other waterfalls of the Lake District, are of the "hanging valley" type. They occur just where tributary streams enter the main valley. The main valley seems to have been deepened by ice action more rapidly than its tributaries, which, therefore, tumble over the steep sides of the main valley as waterfalls. The long-vanished glaciers have left their mark, too, in narrow, saw-toothed ridges, mountain tarns, and deep cirques or corries. The principal tourist centres are **Keswick** on Derwentwater and **Windermere** on the lake of the same name. Both these places can be reached fairly easily by rail, and form good centres for visiting the more distant parts of the district. The inns at the head of Wastwater are most frequented by mountaineers, for the best climbing is to be obtained in that neighbourhood. Apart from catering for tourists, the chief industry of Keswick is the making of lead-pencils. This manufacture affords an example of the principle of **industrial inertia**, or the persistence of an industry after the original advantages have disappeared. Formerly the black lead was obtained from graphite mines in Borrowdale, but these are now worked out. The industry still flourishes by making use of graphite from Ceylon and cedar from Florida.

To speak of the **Pennine Uplands** as a range or a chain is somewhat misleading. They form a hilly mass which is about 160 miles long, and in several places about 30 miles broad. On the whole the district has the character of a plateau, frequently with steep sides and an undulating surface of moorland and bog. In the northern Pennines

AND THE PENNINE UPLANDS 157

the plateau occasionally rises into distinct summits such as **Cross Fell** in Cumberland, and **Whernside, Ingleborough,** and **Penyghent** in west Yorkshire. Whernside and Penyghent consist of outliers of hard Millstone-grit resting on Carboniferous Limestone. The northern part of the Pennines is composed mainly of Carboniferous Limestone with occasional patches of Millstone-grit. The latter formation predominates in the central part, while the **Peak District** in the south is again an elevated area of Mountain

Fig. 60. Dovedale, a valley in a limestone region. Compare with Fig. 55.

Limestone. As in the Lake District, the influence of rock structure on scenery is very marked. In the Millstone-grit region the Pennines form a very distinct, steep-sided plateau. The escarpments of the plateau seem to be crowned by ruined battlements, owing to the mode of weathering of the Millstone-grit. The summit of the plateau consists of extensive, heath-clad moors containing many dangerous bogs.

The scenery of the **limestone areas** is very peculiar. There is hardly any surface drainage, for limestone is easily dissolved by impure water, and the natural joints of the rock are therefore widened, and allow the surface water to escape underground. In places the joints are widened into yawning pits called "swallow-holes," down which the water pours. Many of the streams disappear down these swallow-holes, and, after flowing for a considerable distance in subterranean channels, reappear at lower levels. The solvent action of the subterranean water frequently produces great underground caverns. These are especially well-developed in the Peak District, where they are visited every year by thousands of tourists. The scenery of the limestone uplands is generally somewhat bleak and sterile. At rare intervals, however, river valleys occur, and these are generally of exquisite beauty. Dovedale and Millersdale, in the Peak District, are among the best known examples of these lovely valleys.

Limestone regions are nearly always rich in minerals. We have already seen that the Oolitic Limestone Ridge contains much iron-ore. The fine haematite of the Furness district comes from the Carboniferous Limestone. The Pennines are rich also in lead ores. The Peak District of Derbyshire is one of the most important producers of lead in Britain. The limestone itself is quarried for building stone, or in the form of marble for ornamental work. The soil of limestone districts is generally too thin to be ploughed, but bears a short, sweet grass that nourishes large flocks of sheep.

From what has been said it is obvious that we need not expect to find any large towns on the Pennines. Mining villages and tourist resorts are the chief centres of population. Of the latter, **Matlock** and **Buxton** in the Peak District are the most considerable. Their high situation, their beautiful surroundings, but most of all their mineral

AND THE PENNINE UPLANDS 159

wells and medicinal baths, cause them to be thronged in the season with crowds of people seeking pleasure or health.

The **Pennines** interpose a formidable **barrier to communication** between two busy, industrial districts. On the east are the coalfields of Northumberland and Durham

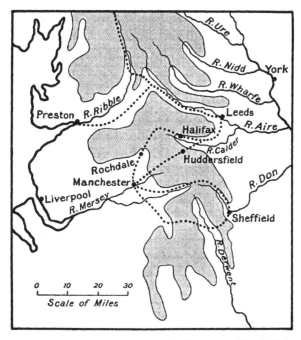

Fig. 61. Railway Routes across the Pennines. The shaded areas are more than 500 feet above sea-level.

and Yorkshire, on the west are the Lancashire and Cumberland coalfields. The Pennines are now crossed by a number of railways, and it will be advisable, therefore, to consider the routes chosen and the reasons for following these routes (see Fig. 61). The Pennine Uplands are trenched right across by two important valleys, the **Tyne**

Gap and the **Aire Gap**. These gaps afford the easiest routes between east and west. The Tyne Gap is followed by the railway between Newcastle and Carlisle. The Aire Gap is utilised by the main Midland line between Leeds and Carlisle. Farther south there are four railway lines that cross the Pennines from the large towns of Lancashire to the great industrial centres of Yorkshire. From Manchester, the Lancashire and Yorkshire Railway runs north-north-east up the Roch valley past Rochdale. It tunnels beneath the highest part of the Pennines, and reaches the Calder Valley at Todmorden. Thence the way is easy down the Calder Valley to Halifax and the other large Yorkshire towns. The London and North-Western line goes north-east from Manchester to Diggle, tunnels the hills for three miles, and then reaches Huddersfield by a tributary of the Calder. The Great Central Railway goes almost due east from Manchester up a tributary of the Mersey. The Pennines are high at this point, and so a long tunnel of three miles is again needed before the eastern slope is reached. The Don Valley then offers the best route to Sheffield. Farthest south is the Midland line. It runs south-east from Manchester, and attacks the plateau where it is divided into two by the north-south valley of the Derwent. Two tunnels are needed, therefore, one to reach the Derwent, the other between the Derwent and a tributary of the Don, whence the line goes north-east to Sheffield.

CHAPTER XXVII

NORTH-WEST ENGLAND

THE district now to be described lies between the Pennine Uplands and the Irish Sea. The northern boundary is formed by the Solway, and the southern by the Midland Gate. The Lake District, rising like an island from the surrounding lower ground of this area, has already been described. The old rocks of the Lake District are found nowhere else in north-west England. We have seen that the Pennine Uplands are composed of lower Carboniferous rocks, either Millstone-grit or Carboniferous Limestone. In Lancashire these rocks are overlaid by the Coal Measures, and still passing westwards we meet the still younger New Red Sandstone. In Cumberland the sequence is the same, although the outcrops are more confused. There are, therefore, in north-west England two very important rock series, the Coal Measures of Cumberland and the Coal Measures of South Lancashire. The Cumberland Coalfield is comparatively small, but the **Lancashire Coalfield** is one of the most important in the country. Not much coal is exported, for this is one of the busiest industrial districts in Britain, and nearly all the fuel is needed for the factories of the neighbourhood. South Lancashire is a hive of industry. Manufacturing towns and villages crowd so closely together that the district is one of the most densely populated in the world. Yet in the midst of it there are two areas that are mostly uninhabited. These two areas lie between Chorley and Bacup. The rocks here are not Coal Measures, but the barren sandstones and shales of the Millstone-grit series. The region is an area of bare moorland, in places 1500 feet above the sea, and is called the South Lancashire Moors. It affords a striking illustration of the control of human activities by the geological structure of the district.

South Lancashire is the home of the **cotton industry**. One could draw a circle of twenty miles radius to include no fewer than eight towns engaged in the cotton trade, all with more than a hundred thousand inhabitants. In addition there are dozens of smaller towns and innumerable villages engaged in some branch of the same industry. The eight largest towns are, in order of size, **Liverpool, Manchester, Salford, Bolton, Oldham, Blackburn, Preston,** and **Burnley.** The raw cotton is imported mainly from the United States, Egypt, and India. The imports for 1910 were as follows:

United States	1470	thousand lbs.
Egypt	329	,, ,,
British India	108	,, ,,
Total from all countries	1973	,, ,,

From these figures it appears that the three countries mentioned supplied in 1910 nearly 97 per cent. of all the raw cotton brought into the British Isles. The imports from the United States comprised exactly three-quarters of the total quantity. Practically all the raw cotton imported into this country comes to Liverpool or Manchester. In 1911 Liverpool imported raw cotton to the value of £53,000,000, while £14,000,000 worth went direct to Manchester by the Ship Canal.

The manufacture of cotton may be considered under four main heads: spinning, weaving, bleaching and dyeing, and the making of suitable machinery. There is a certain amount of specialisation among the cotton towns. For example, Oldham is noted chiefly for spinning, Blackburn for weaving, Bolton for bleaching, and Manchester for the manufacture of machinery. **Manchester** is the centre of the cotton trade. With Salford (for the towns are really

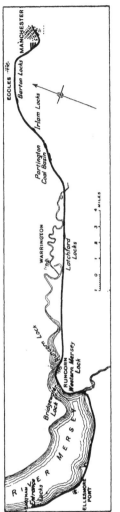

Fig. 62. Sketch-map of the Manchester Ship Canal.

one) it contains a million inhabitants. Apart from machinery there is not so much actual manufacture of cotton in Manchester as in some of the other towns, but the raw cotton is brought here, the finished goods are sold here, and the town is a convenient place for warehousing the cotton, and a **business centre** for the merchants and the manufacturers. Since the opening of the Manchester Ship Canal the city has become a great **seaport**. The tonnage is not so great as in many other ports, but the value of the goods imported and exported is high. In the latter respect it generally ranks fourth among British ports.

CHIEF IMPORTS AND EXPORTS OF MANCHESTER, 1911

Imports	Value	Exports	Value
Raw Cotton	£14,085,000	Cotton goods	£11,881,000
Grain and Flour	3,801,000	Cotton yarn	2,832,000
Timber	1,313,000	Machinery	1,933,000

Consider the position of the towns in Britain engaged in manufacturing cotton. In England, as we have seen, Lancashire is the chief centre. Of the Scottish towns, Glasgow is in west Lanarkshire, Paisley is in west Renfrewshire, and spinning and weaving are also carried on in Ayrshire. The feature common to all these places is that they are on the west coast of Britain. The position of the seats of cotton-manufacture is controlled by a **climatic factor**. A moist climate is desirable. If the air is too dry the threads become brittle and are difficult to work. The west coast of Britain has a much wetter climate than the east, and so, in time, all the cotton-working towns have come to be in the west. A more obvious factor in determining the seats of the industry is, of course, the presence of power-supplies. All the cotton-working towns are on or near coal-fields. As we pass to the south-east of the Lancashire cotton-towns, the climate becomes drier, and there is a corresponding change in the textile that is manufactured. Cotton gives

NORTH-WEST ENGLAND

place to silk. Macclesfield in Cheshire is the chief centre of this trade.

Liverpool is the principal port, not only for the industrial region of south Lancashire, but for many parts of the Midlands as well. As a **seaport** it ranks next to London in importance. Liverpool's trade has always been mainly with **America**. In early times it had a brisk trade with that country in negro slaves, but it was the rise of the

Fig. 63. The Landing-Stage, Liverpool.

English cotton trade and the rapid growth of so many large towns in Lancashire that made Liverpool surpass all its rivals except London. Moored to its twenty-five miles of quays are ships from every corner of the globe. Most of the principal steam-ship lines, including the Cunard and the White Star, have sailings from Liverpool. In former times one of Liverpool's chief rivals was Chester, at the head of the neighbouring estuary of the Dee. But the silting up of the Dee ruined Chester's overseas commerce. Luckily for Liverpool the estuary of the Mersey is bottle-shaped, with the neck seawards, and the scour of the tides through the narrow neck has helped to prevent silting. Liverpool has a considerable Irish trade, but vastly more important is the trade with America. An examination of the imports in the list given below will show that all the commodities mentioned come mainly from North and South America.

CHIEF IMPORTS AND EXPORTS OF LIVERPOOL, 1911

Imports	Value	Exports	Value
Raw Cotton	£52,812,000	Cotton goods	£60,908,000
Grain	14,454,000	Iron and Steel goods	14,713,000
Animals	14,156,000	Woollen goods	9,910,000
Rubber	10,285,000	Machinery	9,862,000
Raw wool	6,921,000	Chemicals	4,407,000
Oil	4,430,000	Cotton yarn	4,000,000
Sugar	4,263,000		
Copper	4,090,000		
Tobacco	3,120,000		

Heysham and **Fleetwood** are other seaports on the Lancashire coast, but chiefly for mail and fast passenger traffic. They have regular sailings to Ireland, and in the summer-time thousands of holiday-makers sail from these ports to the Isle of Man. **Blackpool**, on the Lancashire coast, is one of the largest holiday resorts in Great Britain. The sea-air and the fresh winds form its chief natural attractions, for the Lancashire coast for many miles is low and monotonous, and covered with sand dunes.

On the opposite side of the Mersey from Liverpool is **Birkenhead**, with great engineering works and shipbuilding yards. Farther up the Mersey and also in Cheshire is the chemical manufacturing town of **Runcorn**. It is one of a group of towns engaged in this industry, of which **St Helens** in Lancashire is the largest. The industry is based on the abundant supplies of salt from the New

NORTH-WEST ENGLAND 167

Red Sandstone rocks of Cheshire. Obviously a district which is convenient both to sources of salt and sources of coal is in a favourable position for developing chemical industries, and the towns mentioned fulfil both conditions. St Helens is noted also for its great glass works and its dye factories.

Lancaster is the county town of Lancashire, but it lies over twenty miles away from the coalfield, and thus it has not developed in modern times like the great cotton centres. Its name shows its Roman origin. Like Chester, the

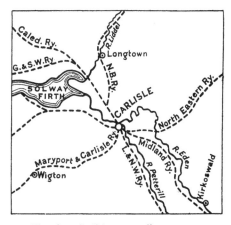

Fig. 64. Carlisle as a railway centre.

Roman station on the Dee, Lancaster was formerly a busy port, but the estuary of the Lune has become silted up, and ships cannot now reach the town. Across the broad waters of Morecambe Bay is **Barrow-in-Furness**, the largest town in Lancashire not on the coalfield. Barrow has had a remarkable history. In the middle of the nineteenth century it was a little fishing village with hardly more than three hundred of a population. At the present time it is a large town with over sixty thousand inhabitants. Its rapid growth is due to the valuable **deposits of iron-ore** that have been discovered in the neighbourhood. The town has many blast-furnaces and steel works, and is noted for its shipbuilding.

North-west of the old rocks of the Lake District, the Coal Measures appear again in the **Cumberland Coalfield**. Many of the underground workings run far out under the sea. The coal is exported from **Whitehaven** and **Maryport**. Iron-smelting is also carried on in these towns, both local and imported ores being used. North of the Lake District the most important town is **Carlisle**. It is one of the best examples in Britain of an old, historic, strongly-fortified town becoming in modern times a **great railway centre**. In the days when England and Scotland were constantly at war, Carlisle was one of the principal strategic centres of the Borders, for it stands at the meeting of several routes. Northwards the way to Scotland lies open. Southwards the Eden valley presents an easy way of reaching the heart of England. Again, just east of Carlisle, the Pennine Uplands are cut through by the Tyne Gap that affords an easy route to the east of England. Plainly a strong castle at Carlisle would keep these important routes safe against an enemy, and this circumstance explains the early importance of Carlisle. In virtue of nodality it became a strategic centre, and in virtue of nodality it became later a railway centre. Three main lines from Scotland come into Carlisle from the north. Two important English railways run south. Eastwards the town is joined to Newcastle by a railway through the Tyne Gap, and westwards a line serves the coast west of the Lake District and the towns of the Cumberland Coalfield. Carlisle was once a seaport with a considerable trade, but the silting up of the Eden has caused it to share the fate of Lancaster and Chester.

CHAPTER XXVIII

NORTH-EAST ENGLAND

IT is convenient to consider **north-east England** as that part of the country that is east of the Pennines and north of the Humber and the Trent. Although the **Coal Measures** have been stripped by denudation from the summit of the Pennines, we saw in the previous chapter that they are found on the western flanks in two patches forming the Lancashire and the Cumberland coalfields. The same structure is found on the eastern side. The Yorks, Derby, and Notts Coalfield is the counterpart of the Lancashire field, while the Northumberland and Durham Coalfield corresponds to the Cumberland field. Passing east from the Pennines we come to younger and younger rocks, first the Coal Measures, then the New Red Sandstone, then the Oolitic Limestone, then the Cretaceous System. These rock systems must be noticed, for they are of considerable importance either economically or as affecting the relief of the area. The economic value of the Coal Measures is obvious. The New Red Sandstone (as in the Midlands) is easily denuded, and therefore forms valleys and plains. The flat **Plain of York** surrounding the cathedral city is composed of New Red Sandstone covered with boulder clay and alluvium. The Oolitic Limestone escarpment and the Chalk escarpment of the Cretaceous System form marked features in this part of England. The **Cleveland Hills** are formed by the outcrop of the Oolitic Limestone, which here swings to the east and reaches the sea in the cliffs between Saltburn and Whitby. North of the Humber the Chalk outcrop forms the **Yorkshire Wolds.** This outcrop also bends to the east and reaches the sea in the cliffs about Flamborough Head.

The **Yorks, Derby, and Notts Coalfield** is the largest in Britain and produces most coal. The towns in the northern part of the field have become the centres of the great **woollen industry** of the West Riding of Yorkshire. In early days the industry began by using the wool of the sheep pastured on the neighbouring hills. The streams rushing down from the Pennines were utilised to drive the mills. It was not until the application of steam to machinery, however, that the West Riding towns, favoured by their situation on the most productive coalfield of the country, attained the commanding position that they now hold. Most of the wool-making centres are situated in the valley of the River Aire. Leeds and Bradford are the largest towns, indeed the former is the sixth largest town in England. In both towns there are huge mills for spinning and weaving the wool. **Leeds** has an enormous trade in ready-made clothing. In recent years the iron and steel works of Leeds have become as important as the woollen mills. In addition to its trade in woollen goods, **Bradford** has large mills for making velvet and plush. **Halifax** and **Huddersfield** are two other large towns with over a hundred thousand inhabitants. Both specialise in certain branches of the woollen trade. Halifax makes more carpets than any other town in the world, while the fine broadcloth of Huddersfield is famous.

It is many a year since the Yorkshire woollen industry outgrew the home supplies of raw material. The **imports of raw wool** into this country come chiefly from the British Colonies. Australia, New Zealand, and South Africa are by far the most important in this respect. In 1911 our imports from these countries were as follows:

IMPORTS OF RAW WOOL, 1911

Australia	323 million lbs.
New Zealand...	174 ,, ,,
South Africa...	101 ,, ,,

Of the countries outside the British Empire the most important source of our wool supply is the Argentine Republic. The South American wool comes mainly to Liverpool. The wool from Australasia comes to London and is thence distributed. About two-thirds of the total imports come to the latter port.

In the extreme south of Yorkshire, but still on the coalfield, is the great steel town of **Sheffield**, the fifth city in England for size. For many centuries Sheffield has been noted for its **cutlery**. Even Chaucer speaks of a "Sheffield whittle," the knife carried by the common people of his time. It is probable that the cutlery trade began in Sheffield because of the supply of coarse sandstones, suitable for grindstones, found in the neighbourhood, and also because of the water power that was at hand. Everything that can be made of **steel** is made in Sheffield—armourplate for battleships, steel rails, locomotives, big guns, motor cars, and machinery of all kinds. The main line of the Midland Railway passes through Sheffield and Leeds.

North-east of the coalfield stretches the flat Plain of York. It is an agricultural district, and its chief town is **York**, situated in the centre of the Plain, and on the principal river of the district, the Yorkshire Ouse. York, therefore, possesses a high degree of nodality, and so has become the county town and the chief market town for this region. The advantages of situation possessed by York were recognised long ago by the Romans, who had one of their chief settlements there, namely, Eboracum, from which the modern name is derived. The principal attraction of the town, nowadays, is its magnificent cathedral, York Minster. The fact that York is a meeting place of many routes has made it a railway centre of considerable importance.

Hull, on the Humber, is much the largest **seaport** in north-eastern England. It ranks after Liverpool, taking third place among British ports. It has a large trade with all the countries of western Europe, to which it exports the woollen goods of Yorkshire and the cotton goods of Lancashire. Like Grimsby it is a convenient port for the North Sea fishing-banks, and so it ranks second only to that town in the **fish trade**. Indeed Hull and Grimsby together receive as much fish as all the other ports of England put together. Hull imports large quantities of **oil-seeds**, and is the chief centre in Britain for crushing the seeds in order to procure the oil. The following table gives the chief imports and exports of Hull for 1911.

CHIEF IMPORTS AND EXPORTS OF HULL, 1911

Imports	Value	Exports	Value
Grain	£10,144,000	Cotton goods and yarn	£4,775,000
Seeds (chiefly oil)	6,235,000	Machinery	3,544,000
Butter	3,176,000	Woollen goods and yarn	2,589,000
Wool and woollen goods	1,899,000	Coal	1,821,000
Timber	1,870,000	Iron and Steel goods	1,580,000
Meat	1,554,000	Oil	1,244,000
Eggs	1,517,000		

The scenery along the Yorkshire coast is very fine, and the climate is dry and bracing. Many of the coast towns are therefore favourite summer resorts. **Whitby** is one of the most interesting of them all. It is a quaint, old, fishing town with steep and narrow streets. The antiquity of the town is witnessed by the fact that a famous council was held there as long ago as the seventh century. The abbey is one of the most beautiful, old churches in England. **Scarborough**, a little farther south, is a more fashionable watering-place. It has a lovely situation at the head of a sandy bay encircled by steep cliffs, which rise north and south of the town into fine headlands. Although far inland, in the basin of the Nidd, **Harrogate** may be mentioned here, for it is the principal inland watering-place of north-east England. Its mineral springs have a high reputation,

NORTH-EAST ENGLAND

and people come from all over the country to take the waters.

In the north-east of England is found another of the great coalfields of this country. This is the **Northumberland and Durham Coalfield**, the output of which is surpassed only by that of the Yorks, Derby, and Notts Coalfield. Great quantities are exported to foreign countries, and there is also a particularly big coast-wise trade in coal. As in all the coalfields of this country we find mining and engineering as the fundamental industries.

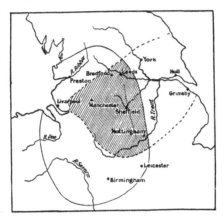

Fig. 65. Overlapping of the hinterlands of Liverpool and Hull. The shaded area is served by both ports. (After Chisholm.)

But most of the coalfields are associated with an additional industry. Cotton-working is characteristic of the Lancashire field, metal-working of the Black Country, woollen-working of the Yorkshire field, pottery-making of the North Staffordshire field, and so on. In Northumberland and Durham, in addition to coal-mining and engineering, we find a great **shipbuilding** industry. The banks of the Tyne, the Wear, and the Tees are lined with big shipbuilding yards. The Tyne is the most important shipbuilding river in England, and is only surpassed in the world by the Clyde. The actual tonnage produced in 1912 is shown on p. 86. Obviously the advantages for this industry are similar to those already discussed in connection with the Clyde.

The chief town on the coalfield is **Newcastle**, on the River Tyne. Its coal trade is of course proverbial. There are large engineering works where big guns, locomotives, engines for ships, and machinery of all kinds are made. The chemical and glass industries of Newcastle and district have a basis similar to that of the same industries in South Lancashire already discussed. The New Red Sandstone rocks of the Tees basin are rich in salt and other minerals useful to the manufacturing chemist. The other towns on the Tyne, **Gateshead**, **North Shields**, **South Shields** and **Tynemouth** have industries similar to Newcastle but on a smaller scale. The following are the chief imports and exports of Newcastle, including North Shields and South Shields.

CHIEF IMPORTS AND EXPORTS OF NEWCASTLE, N. SHIELDS AND S. SHIELDS, 1911

Imports	Value	Exports	Value
Butter	£2,413,000	Coal	£6,147,000
Grain and flour ...	1,294,000	Ships	1,336,000
Timber	752,000		
Meat	704,000		

Sunderland, at the mouth of the Wear, is a great shipbuilding centre. Higher up the river is **Durham**, which has a university, and a fine, old cathedral. **Darlington** and **Stockton** are the principal towns on the Tees. Their trade is the same as that of the other towns of this district. Coal is exported, ships are built on the banks of the Tees, machinery is made, and locomotives are built. The last-named industry is certainly appropriate to this district for

NORTH-EAST ENGLAND 175

George Stephenson was born near Newcastle. This is the richest iron-producing area in this country. The Cleveland Hills, of course, form the termination of the Oolitic Ridge,

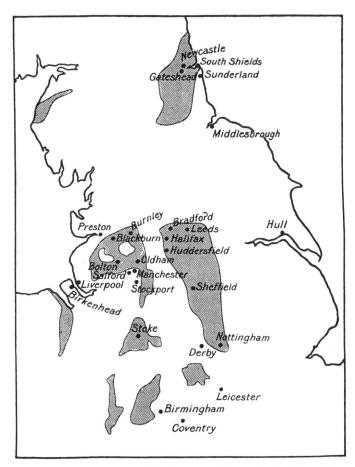

Fig. 66. The large towns of Northern and Midland England in relation to the Coalfields. All the towns shown contain over 100,000 inhabitants.

which, we have already seen, is rich in iron-ore all along its course. The chief town of the Cleveland district is

Middlesbrough which has a history very similar to that of Barrow. Less than a century ago there was only one house where now there is a great town with more than a hundred thousand inhabitants. Middlesbrough is now the most important **iron** town in Britain.

From what has been said in the last few chapters we may conclude that in this country the most powerful factor in the growth of towns has been **proximity to coal.** Fig. 66 illustrates this point in a striking manner. It shows the coalfields of north and midland England, and all towns with over 100,000 inhabitants at the 1911 census. Of the twenty-seven towns shown, over twenty are either on or just at the edge of a coalfield.

CHAPTER XXIX

WALES

Wales is the most mountainous part of the British Isles south of the Scottish Highlands. The Welsh Mountains consist of a roughly rectangular block of old rocks, which has been carved by rain, rivers, frost, and ice, into a picturesque region of mountains, lakes, and valleys. In the north **Snowdon** is nearly 3600 feet high, **Plynlimmon** in the centre is a thousand feet lower, while in the south the **Brecknock Beacons** rise again to 2900 feet. Wales, therefore, resembles the Scottish Highlands in being a dissected plateau which has been carved from a pene-plain now elevated about 3000 feet above the sea. There are traces of another base level of erosion about 1200 or 1400 feet above the present level of the sea. Before the advent of the Ice Age this base level was quite distinct, and

WALES

appeared as an upland plain from which the mountains rose in smooth slopes with rounded summits. Sharp peaks and serrated ridges were absent, there were few or no lakes, and the tributary streams entered the main river at accordant junctions. Much of the picturesqueness that characterises Wales to-day was produced during the Ice Age. The rounded mountains were sculptured into narrow, serrated ridges separating great cirques or cwms (kooms) as they are called in Wales. Lakes were formed either by erosion

Fig. 67. A distant view of Snowdon.

of rock basins or the damming of valleys by glacial deposits. The main valleys were deepened more than their tributaries, and the latter therefore appear now as "hanging valleys," which join the main streams by cascading over the steep sides of the main valleys. All these features can be seen to perfection in the district round Snowdon. The scenery there is wilder, more rugged, and on a grander scale than in any other part of Britain outside the Scottish Highlands.

For the most part Wales consists of a complex of **ancient rocks**. Where these are penetrated by igneous rocks, a very hard and resistant rock series is formed which stands above the general level, as in the district round Snowdon, or forms prominent headlands such as Braich-y-Pwll or St David's Head. Towards the south the grits and slates of the Cambrian and Silurian systems are replaced by Old Red Sandstone rocks, which form the Brecknock Beacons and the Black Mountains of Brecknockshire. The extreme south of Wales is in geological structure a great trough containing the valuable **Coal Measures of South Wales**. The long axis of the trough runs east and west, and fragments of the broken southern rim are found in the extreme south of Pembroke, the Gower peninsula of Glamorgan, and the extreme south of the latter county.

The **rainfall** over parts of Wales is excessively high. The district round Snowdon is one of the rainiest parts of Britain, and one of the few districts of Europe with a rainfall of more than a hundred inches a year. This fact has been taken advantage of by some of the large cities of England, which have built **reservoirs in Wales** in order to secure an abundant supply of pure water. The artificial Lake Vyrnwy, now the largest lake in Wales, is the Liverpool water-works. It was an ancient lake basin in the extreme upper valley of the Severn which has now been refilled. Among the mountains south of Plynlimmon there are two other artificial lakes These were made by damming a tributary of the Wye, and form the water-works of Birmingham.

From what has been already said it is plain that in many respects Wales bears a strong **resemblance to the Lake District**. Both regions are mountainous districts, with fine scenery that attracts thousands of tourists in summer time. The type of scenery, too, is similar in both

WALES

districts, for many of the characteristic features are due to ice sculpture. The rocks are approximately of the same age and are similar in character. The rainfall is very high, and both districts are therefore used as water-catchment areas for distant cities. Since the physical characters of the two regions are so similar, we might expect similar industries, for economic conditions have all a physical basis. It is not surprising, therefore, to find in both the Lake District and Wales (excluding South Wales) that

Fig. 68. Carnarvon Castle.

sheep-rearing and slate-quarrying are among the most important industries.

In some respects, too, Wales is **similar to Scotland**, but there are also striking differences, and both the resemblances and the contrasts in physical structure are reflected in the historical parallels and differences between the two countries. Both countries, being mountainous, became the refuge of the Celtic-speaking races who were driven there by their enemies the Anglo-Saxons. But in the heart of Scotland there is the Lowland Plain which encouraged a settled population, and made possible a united Scotland under one ruler. Wales never became a single kingdom. There is no real centre in Wales to form the heart of a single, united people. The country falls into three natural regions, Snowdonia or North Wales, most inaccessible and, therefore, last to be subdued, Central Wales, open to attack by the Severn valley, and South Wales, more mountainous than the centre but not so rugged as Snowdonia. These three divisions correspond to the ancient principalities of Gwynedd, Powys, and Dinefawr. The Welsh people were therefore ruled by several independent chiefs instead of one king as in Scotland. In the Wars of Independence waged both by Scotland and by Wales against England, the former country was more successful than the latter, not from any less determination or inferior valour on the part of the Welsh, but because their geographical environment was less favourable. The most striking relics of the English wars of aggression are the splendid, old Norman castles. Perhaps the finest of these is **Carnarvon Castle**, one of the most magnificent ruins in Britain. Many of the great towers are still in fine condition. The castle was founded by Edward I, and in it the first Prince of Wales was born. Another splendid ruin is **Conway Castle**, also built by Edward I. There are eight vast towers, and the walls in places are fifteen feet thick.

The **coalfield of South Wales** is one of the great industrial regions of this country. Although barely as productive as the Yorkshire or the Northumberland field, its export trade is the greatest of any field in the world. This is partly accounted for by the peculiar suitability of the Welsh coal for steamships, for much of it is smokeless. In the east of the field the coal is the ordinary bituminous variety. In the centre it is harder and has a higher percentage of carbon; it is now an excellent smokeless or "steam" coal. In the extreme west a pure anthracite is found. The Government owns mines in the central part of

WALES

the field for the supply of coal to our navy. **Cardiff**, at the mouth of the Taff, is the great **coal-exporting centre**. It sends abroad more coal than any other town in the world. The coalfield is drained by rivers flowing to the Bristol Channel, and down the valleys, day and night, by railway and canal, there pours a stream of trucks and barges, bringing coal to the great docks at Cardiff. Cardiff actually handles a larger tonnage of shipping than any other port in Britain, except London and Liverpool, although of course the value is not nearly so high as that of several other ports. The town contains great blast

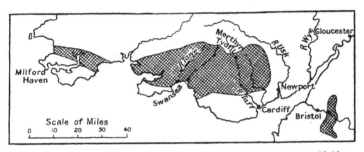

Fig. 69. Sketch-map of South Wales. The shaded areas are coalfields.

furnaces for the smelting of iron. Newport, at the mouth of the Usk, is another large coal-exporting town.

CHIEF IMPORTS AND EXPORTS OF CARDIFF, 1911

Imports	Value	Exports	Value
Grain	£2,684,000	Coal	£11,887,000
Timber	1,393,000	Iron and Steel goods	382,000
Iron Ore	712,000		

Swansea, at the mouth of the Tawe, is the most important **metal-smelting** town in Britain. Large quantities of copper are refined, and it is the greatest centre in the world of the "**tin-plate**" industry. A large number of kitchen utensils are made of tin-plate, as well as cans and tins of every kind. These articles are made from thin sheets of iron that have been dipped in molten tin. Swansea is an excellent example of the principle of "**industrial inertia**" that we have already seen exemplified in Keswick and other towns. The South Wales coalfield is the nearest large field to Cornwall and Devon, where in former times all our copper and tin were obtained. These ores were therefore carried across the Bristol Channel to Swansea and smelted there. The mines of Cornwall and Devon are now largely exhausted, and we obtain most of our tin from the East Indies, and most of our copper from Spain. But in spite of the disappearance of the original geographical advantages, Swansea still holds its position in virtue of the perfected organisation, the expended capital, and the skill of the workers, which give an "inertia" sufficient to counterbalance the lost advantages of position.

The same principle is illustrated by the industrial history of the northern part of the coalfield. This is a great **iron-smelting** district, **Merthyr Tydfil** in the upper Taff valley being the chief centre of the industry. Originally the iron-ore was obtained from local deposits, but these are now largely worked out, and large quantities of ore are imported from Spain. In spite of the fact that such a heavy material as iron-ore has to be sent by rail to the northern part of the field, the industry still flourishes.

Outside the southern coalfield there are no large industrial towns in Wales. Again, except in some of the valleys or the coastal plain, most forms of agriculture are carried on at a disadvantage. But **sheep-rearing** is a very important industry. For its size Wales possesses more sheep than either England, Scotland, or Ireland. One expects, therefore, to find woollen manufactures, and this is the case. Owing to the absence of coal, however, in the hilly districts where the sheep are reared, this branch of industry has never grown to great importance, although Welsh flannels are famous. The woollen trade is carried on chiefly in

WALES

Montgomery. The low-lying island of Anglesey is a famous cattle-rearing district. The slates of Wales are the finest that can be obtained in this country. In the north of Carnarvon are the enormous slate quarries of Bethesda, the largest in Britain. Looking across the valley, too, from the north slope of Snowdon, one sees the sides of the opposite mountains gashed and cleft by the hands of the quarrymen.

The beautiful scenery of Wales makes the country a favourite place for holiday-makers. All round the coast there are to be found little towns that have grown rapidly in recent years as holiday resorts. On the north coast **Llandudno** is the chief watering-place. It has a magnificent bay with fine headlands on each side, while in the background rise the rugged peaks round Snowdon. On the west coast **Aberystwith** is the most popular resort, while some people think the scenery about **Barmouth** is the finest in the country. Two of the coast towns are important packet-stations for Ireland. **Holyhead**, on Holyhead Island off Anglesey, is just sixty miles from Dublin, and is connected with the latter town by a service of swift steamers. **Fishguard**, in Pembroke, is just over sixty miles from Rosslare, the port of Wexford, and speedy cross-channel steamers ply between the two ports. Some of the American mails, too, are landed at Fishguard, and sent thence to London by express train. In the extreme south-west of Pembroke is the magnificent natural harbour of **Milford Haven**, which shelters the government docks near the town of Pembroke. On the opposite side of the estuary is Milford at which great quantities of fish are landed. Most of the Irish fishing-boats discharge their catches here. In some years Milford is surpassed only by Grimsby in the amount of fish landed.

Many of the **place-names of Wales** are similar to those we have already encountered in the Scottish Highlands.

The Welsh language, like the Gaelic, belongs to the Celtic group. But the Celtic races are divided into the Goidels and the Brythons, and Gaelic is a Goidelic tongue while Welsh is Brythonic. A few examples will illustrate the point. *Aber*, the mouth of a river, is the same in both languages; the Gaelic *ben*, a mountain, becomes the Welsh *pen*; *more*, big, in Gaelic, is *mawr* in Welsh; *beg*, little, becomes *bach* in Welsh; the Gaelic *sron*, a nose or cape, changes to the Welsh *trwn* (pronounced troon). The commonest place-name in Wales is *llan*, a church. Another common prefix is *caer*, a castle, and *pont*, a bridge, is also frequently found.

CHAPTER XXX

RAILWAYS AND COUNTIES

THE main **lines of communication** in this country depend in the first place on the fact that there exist widely separated centres of population between which it is necessary that there should be intercourse. The actual routes chosen by the railways afford in many cases good illustrations of the way in which human activities are controlled by physical conditions. For this purpose railways offer a much better commentary on the relief than roads, for locomotives are not good climbers, and easy routes must be found. While a gradient of 1 in 40 is a stiff hill for a railway, many roads have hills as steep as 1 in 10, and there are some roads in Britain with a gradient of 1 in 5, although this is excessively steep. The average cyclist dismounts at a hill of about 1 in 17. We shall describe in detail two of the main lines from London to

Carlisle. The other lines mentioned should be traced on the map, and an attempt made to find reasons for the particular courses taken by the railways.

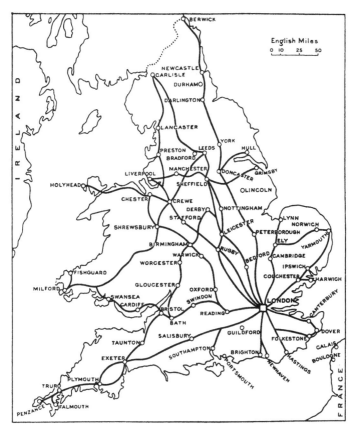

Fig. 70. The Main Railway Lines of England.

The **London and North-Western Railway,** as its name shows, connects London with Lancashire and the north-west of England. It starts from Euston Station and goes north-west. Now it will be remembered that there are two lines of hills that it must cross, first the chalk ridge of the Chiltern Hills, and next the limestone ridge of the Cotteswolds. The railway will cross these hills by the easiest route that will not take it too far out of its way. The Chilterns are cut through by a number of dry valleys, and the railway makes use of one of them at Berkhampstead in order to cross this barrier. There is no convenient valley cutting right through the limestone ridge, and therefore this obstacle is passed by a tunnel near **Rugby**. The railway next runs through **Stafford** and **Crewe**, in the broad, flat passage known as the Midland Gate, and is soon on the western side of the Pennine Uplands, having come round the southern end of this formidable barrier to railways. The line now goes through Lancashire by way of **Wigan, Preston,** and **Lancaster.** Carlisle is the town we are trying to reach, but the way there is barred by hills. The hilly region is lowest and narrowest where the Lake District mountains are joined to the Pennines by a narrow neck of high ground. Now up one side of the neck there is an easy route by the Lune valley, and on the north side there is the valley of the Petterill, a tributary of the Eden. The railway therefore climbs up the Lune valley, crosses the neck of high ground at **Shap Summit**, 1000 feet above sea-level, and runs down the Petterill valley to **Carlisle.** At Carlisle the London and North-Western Railway joins the Caledonian Railway of Scotland.

The Pennine Uplands separate two densely populated districts, the cotton towns of Lancashire on the west from the woollen towns of Yorkshire on the east. As we have seen, the London and North-Western Railway serves the Lancashire towns; on the other hand the Yorkshire towns are connected with London by the **Midland Railway.** It starts from St Pancras Station, and since it intends to keep to the east of the Pennines, its course soon diverges from that of the London and North-Western Railway, and keeps more

RAILWAYS AND COUNTIES 187

to the east. In the Midlands, **Leicester, Nottingham,** and **Derby** are the chief towns passed, and then the line enters Yorkshire and reaches **Sheffield.** After passing **Leeds** the railway has a difficult task before it, for it has to reach Carlisle, and the Pennine Uplands bar the way. But the **Aire Gap**, opening west from Leeds, offers a likely route. The railway therefore uses this gap, and then climbs the Pennine Uplands by means of the Ribble valley. It crosses the lonely moors that form the top of the Pennines, reaching a height of 1200 feet, and then runs down the long valley of the Eden, which brings the line into **Carlisle.** At this town the Midland Railway connects with the Glasgow and South-Western Railway of Scotland.

In addition to Yorkshire and Lancashire there is still another busy part of northern England, namely, the Northumberland and Durham Coalfield. Newcastle and the other large towns in this neighbourhood are connected with London by the **Great Northern Railway** (London to York), and the **North-Eastern Railway** (York to Berwick). These two lines are often called by the one name of the East Coast Route, just as the London and North-Western line is known as the West Coast Route. The **Great Western Railway** has the greatest mileage in Britain, and, as its name indicates, serves chiefly the west coast of England from Cornwall to North Wales. One important branch runs through Oxford, Birmingham, Shrewsbury, and Chester, to Birkenhead. Other branches connect Bristol, Cardiff, and Swansea with London. Exeter, Plymouth, and Penzance are on another branch of this line. Exeter and Plymouth are also served by the **London and South-Western Railway,** which sends another branch to the great port of Southampton. The **Great Central Railway** connects London with the Midlands, and also sends branches to Lancashire, Yorkshire, and other northern counties. The railway lines that serve the seaports of the

English Channel are of special importance. Dover and Folkestone are connected with London by the **South-Eastern and Chatham Railway,** while Harwich and Yarmouth are served by the **Great Eastern Railway.** Brighton and the towns on the south coast communicate with London by means of the **London, Brighton, and South Coast Railway.**

Let us now consider very briefly the **origin of some of the counties** of England. It would be quite impossible to discuss every county in England, therefore we shall select a few instances in which history has been shaped by geographical conditions, paying special attention to county boundaries, for in many cases they have been formed by marked geographical features. The south-east of England was conquered by the Angles, Saxons, and Jutes. A settlement was made along the coast, and their outposts were gradually pushed inland. Any marked physical barrier would probably form a boundary between different tribes. For example, the dense forests of the Weald formed the boundary between Sussex and the counties of Surrey and Kent to the north. Where the eastern boundary of Sussex leaves the Weald and turns south-east, there was a line of marshes ending at Romney. In this region at the present time the most powerful natural barriers are the chalk escarpments of the North Downs and the South Downs. The Weald no longer prevents communication between north and south. We must in imagination restore its forests in order to perceive its power of checking an advancing army.

The Thames, unlike all the other great rivers of England, forms a boundary between two sets of counties. This is usually the case only with the lower part of a river, where its width and depth make it a formidable barrier. The Severn and the Yorkshire Ouse illustrate the usual part played by a river. It was the presence of London that made history take a different course in the case of the Thames. London held out for a long time against the invaders, and so barred the advance inland by the Thames. The invading tribes therefore pushed their way to the Thames from their settlements in the south, and so the river became a boundary. The Lea and the Colne flow southwards to the Thames through broad marshes, and thus we find these rivers forming the east and the west boundaries of Middlesex. The western boundary of Norfolk, Suffolk, and Essex was formed by the natural barriers of the Wash, the Fens, and the wooded escarpment of the Chilterns.

In the adjustment of county boundaries the Yorkshire Ouse played a very different part from the Thames. It offered an easy route to the invaders, and fortified positions on it became the centres of settlements that naturally embraced territories on both banks of the river. The basin of the Ouse became the Anglian Kingdom of Deira, with its capital at York, which is situated where the river flows through a wide and fertile plain. Where the Ouse widens into the Humber estuary, it forms a boundary between the counties of York and Lincoln. Most of the northern boundary of Yorkshire is formed by the Tees, which thus resembles the Thames in being a line of separation. The causes at work were not unlike those operating in the case of the Thames. The Angles of Deira pushed their frontiers northwards from the Plain of York. But the north of England constituted the rival Anglian Kingdom of Bernicia, the territory of which was being widened by advances southwards from the capital at Bamburgh. The Tees valley therefore became the frontier between the kingdoms, and later the boundary between two counties.

Nearly all the Midland counties belong to one type. It will be noticed that the names of these shires are the same as the names of the county towns, in which respect they differ from such counties as Kent, Surrey, and Sussex. The Midland counties crystallised into separate entities on the recapture of this district from the Danes. A strong town, usually on a river, was seized and fortified, and became the centre from which English influence radiated. The counties therefore formed around the chief town of the district, and on the county being handed over to one of the nobles for administration, his headquarters were made at the county town. Since the county towns are usually on the chief rivers of the district, we do not expect these rivers to play a prominent part as county boundaries. Most of the Welsh counties are plainly of the Midland type.

CHAPTER XXXI

THE STRUCTURE OF IRELAND

IN the first chapter of this book we gave reasons for thinking that Ireland was formerly joined to Great Britain, but that in comparatively recent times (geologically speaking), the land between them had foundered, and left Ireland as a separate island. If this be the case, then we should expect to find the rock structures of Great Britain repeated across the Irish Sea. An examination of the **structure of Ireland** shows that this is the case. The rocks of Ireland are the westward prolongation of the rocks found in Scotland, England, and Wales.

The Highlands of Scotland consist of a complex of hard, metamorphic rocks, much broken on their western border, and traversed by fractures and lines of weakness, the most dominant of which run from north-east to south-west. Exactly the same rocks are found in the north of

THE STRUCTURE OF IRELAND

Ireland. Donegal, Tyrone, Londonderry, and northern Antrim are composed of very **ancient schists and gneisses**, although in Antrim they are largely covered by a thick sheet of younger volcanic rocks. Similar old rocks are found again in western Mayo and Galway. These ancient crystalline rocks are injected by granite, and run in ridges with a trend from north-east to south-west. The resistance of these rocks to weathering gives rise to the strongly-marked relief, the fine scenery, and (unfortunately) the barren character of much of Donegal, Mayo, and Galway.

The **volcanic rocks** of Antrim form a high plateau that fronts the sea in lofty cliffs. These lavas are exactly similar to the volcanic rocks of the Inner Hebrides of Scotland. There is little doubt that in former times a huge sheet of lava stretched continuously from Skye to Ireland, but, on the breaking of the land bridge between Scotland and Ireland, much of this area sank beneath the waves of the Atlantic Ocean. Parts of the **Antrim Plateau** rise to a height of nearly 2000 feet above the sea. The coast scenery round this part of Ireland is magnificent, but the interior is somewhat bleak and monotonous.

The Southern Uplands of Scotland have their counterpart in Ireland. The **Silurian rocks** of Wigtownshire reappear in the old rocks of County Down. These rocks can be traced in a south-westerly direction, until in the very middle of the country they dip under the younger rocks of the Central Plain. In Ireland, as in Scotland, the Silurian rocks are pierced by **granite bosses**, which in the **Mourne Mountains** rise nearly 3000 feet above the sea. These granite bosses are referred to the same period (early Tertiary) of volcanic activity as produced the lava sheets of Antrim, but while the latter were surface flows from great cracks, the granite probably represents the subterranean reservoirs of molten rock now laid bare by the denudation of the overlying strata.

Most of the centre of Ireland is occupied by **Carboniferous rocks** which stretch without a break from Dublin Bay to Galway Bay. Unfortunately the most valuable rocks of the Carboniferous System, namely, the Coal Measures, have been worn away, and, therefore, most of the manufactures of Ireland depend on imported coal. Most of the Central Plain consists of Carboniferous Limestone. In a few places, Sligo and Clare for example, the limestone forms scarped hills like those of Yorkshire or Derbyshire, and the scenery is of the peculiar type associated with limestone uplands, and described in the chapter on the Pennines. But for the most part the limestone area is lowlying, water-logged, and quite unpicturesque. The rivers of the Central Plain dissolve the limestone, and expand into shallow lakes which are very characteristic of a waterlogged limestone area. In many parts ancient lakes have been filled with moss and changed into bogs, of which the **Bog of Allen** is the largest. The peat from the bogs forms a valuable fuel, and is also used for horse litter. Many attempts have been made to convert Irish peat into a fuel that can compete with coal, but hitherto the results have been disappointing. In a few places isolated hill masses rise suddenly from the Central Plain like islands from the sea. These mountains mark where the resistant older rocks pierce through the cover of Carboniferous Limestone. The **Slieve Bloom Mountains** and the other hill masses round Lough Derg are formed of hard Silurian and Old Red Sandstone rocks.

The **Wicklow Mountains** attain over 3000 feet above the sea at Lugnaquilla. This is one of the most beautiful districts in Ireland. **Glendalough**, in the heart of the mountains, is considered by some the finest valley in Ireland. Between the mountains and the sea the lovely **Vale of Ovoca** has been immortalised by Moore. The old rocks of Wicklow and Wexford resemble in most

THE STRUCTURE OF IRELAND 193

respects the Welsh massif on the opposite side of the Irish Sea, and in former times these two areas undoubtedly formed one continuous highland. Like the Mourne Mountains farther north, many of the Wicklow Mountains are composed of granite.

The south-west of Ireland is the most mountainous part of the country. The **Kerry Mountains** are the highest in Ireland, Carrantuohill, the loftiest summit, attaining a

Fig. 71. Glendalough.

height of 3400 feet, that is, it is higher than any mountain in England, but barely so high as Snowdon. Farther east the **Galty Mountains** also surpass 3000 feet in height. The beautiful lake district of **Killarney,** overlooked by the summits of the Kerry Mountains, is of all parts of Ireland the most visited by tourists. It is plain from the map that the mountains of south-west Ireland run in parallel ridges that have an east and west direction. For the most part these ridges belong to the Old Red Sandstone system, with occasional patches of more ancient rocks. These ridges are the remains of a great mountain system (the Armorican Mountains) that formerly stretched from Belgium across Brittany, England, and Ireland, and far west into the Atlantic Ocean. They have received their name from the fact that they were highest in Brittany, the ancient Armorica. These mountains were formed at the close of the Carboniferous epoch by tremendous pressure acting in a north and south direction. This Armorican uplift is of fundamental importance in the physical history of southwest Ireland, for it has determined the whole character of the area.

The **directions of the rivers** of southern Ireland have been profoundly modified by the structure of the country. When the rivers originated, the Armorican ridges were probably buried under younger rocks which formed a plateau with a gentle tilt towards the south or south-east. The consequent rivers developed on this plateau therefore flowed towards the south or south-east. When the younger rocks were worn away, subsequent rivers developed, and, as the Armorican ridges became etched into greater and greater relief, the subsequent rivers in the softer troughs were plainly working under more favourable conditions than the consequent rivers that had to saw down their valleys across the hard ridges. The subsequent streams therefore pushed their headwaters rapidly towards the west, and beheaded several of the consequent rivers. Thus the Suir probably decapitated the consequent part of the Blackwater, the Blackwater beheaded the Lee, and the Lee beheaded the Bandon. All these rivers are now represented by a long, eastward flowing, subsequent course, and a short, southerly, consequent course at their mouths. Farther east, the Slaney and the Barrow, which now cut right through the heart of the granite mountains of Wicklow, must have originated on the surface of a vanished

THE STRUCTURE OF IRELAND

plateau, the level of which rose above the tops of the highest mountains of to-day.

Summing up, therefore, we have seen that the highlands of Ireland occur in detached masses near the coast, while the centre of the country is occupied by a great plain. Intercourse is therefore much easier than in Scotland, where the highlands occur in larger masses, and thus constitute more formidable barriers to communication. The relief of Ireland is seen to be a result of its rock structure. The older rocks of Great Britain find their counterparts across the Irish Sea, and these rocks by their superior resistance to the weather now stand out in bold relief as highland areas.

CHAPTER XXXII

THE CLIMATE, AGRICULTURE, AND INDUSTRIES OF IRELAND

ALTHOUGH the mountains of Ireland are found round the rim of the country, a glance at an orographical map shows that these highlands cannot form a climatic barrier. They occur in detached groups, leaving between them wide gaps that give the rain-bearing winds from the sea easy access to the heart of the country. The **climate** of Ireland might be summed up as **mild and moist**. The temperature is equable, and rainfall occurs fairly evenly over the whole country. The mean annual range of temperature of south-west Ireland is less than that of any other part of the British Isles. In this respect a comparison between Valentia and London is instructive.

Place	Mean January Temperature	Mean July Temperature	Mean Annual Range
Valentia	42° F.	59° F.	17° F.
London	38° F.	64° F.	26° F.

The **rainfall**, too, is evenly distributed and fairly abundant. The average for the whole country is probably a little over forty inches per annum. Only in the highlands of Mayo and Kerry does the rainfall rise above sixty inches, and only in a small area round Dublin is it less than thirty inches. The general dampness of the climate is no doubt one of the causes of the high death-rate from consumption in Ireland.

Ireland gets its name of the Emerald Isle from the rich, green pasture-lands that cover so much of its surface. In Ireland, as in other parts of the British Isles, the tendency in recent years has been to put more of the land under **pasture** at the expense of grain and other crops. At the present time there are roughly 10 million acres under permanent pasture, and 2⅓ million acres under grain and other crops. **Oats** are the chief grain crop, barley and wheat taking a very subsidiary position. The great preponderance of oats is obviously due largely to climatic causes. The areas under the chief grain crops in 1910 were as follows:

            Oats   ...   ...   1074 thousand acres
            Barley   168   ,,   ,,
            Wheat   48   ,,   ,,

Putting it another way, we may say that Ireland produces nearly a third of the oats grown in the United Kingdom, one-tenth of the barley, and only one thirty-seventh of the wheat. **Potatoes**, however, are widely grown. More of this crop is produced than in either Scotland or England At one time **flax** was grown almost everywhere throughout the United Kingdom, but the plant is now confined to north-east Ireland. About a quarter of the flax used in the great linen industry of Ulster is home grown. The amount produced, however, is steadily shrinking. Thirty years ago the acreage under flax was more than double what it is now. One of the causes of the diminution of the crop is that the preparation of the fibre needs a considerable amount of hand labour, and this can be procured more cheaply in other countries.

The rearing of **live stock** is by far the most important industry in Ireland. Although Scotland and Ireland are approximately equal in area, the latter country possesses four times as many cattle as the former, twice as many horses, and actually ten times as many pigs. With sheep, however, the positions are reversed, for Irish pasture land is largely of a rich, lowland type, while Scottish grazing districts are largely upland. Dairy farming is one of the chief occupations. Great quantities of butter, cheese, and eggs are sent to Great Britain from Irish ports. Large numbers of cattle, too, are reared in order to be exported as food. Among the meat products, Irish-cured bacon takes an important place.

The **fisheries** of Ireland are not so valuable as those of Scotland or England, but they are more important than would appear from Board of Trade returns, for a large proportion of the catches of Irish boats is never landed at home ports, but is sent direct to England. The west coast of Ireland is the most unprotected in the British Isles, and, therefore, while fishing is carried on from almost all the seaport towns of the west, the people are unable to fish continuously, and thus a race of crofter-fishermen is characteristic of the west coast. The inhabitants who live solely by fishing are found along the east coast, where Arklow is the chief fishing centre. The fishermen of the south-west bring their catches to Kinsale, Valentia, and similar centres, whence they are taken in steamers to Milford. Considerable quantities of fish are salted and exported to America.

Ireland is handicapped as a manufacturing country owing to the **scarcity of coal**. The Coal Measures occur in patches resting on the Lower Carboniferous rocks, and are obviously the remnants of an extensive sheet now mainly removed by denudation. In Antrim, Kilkenny, and other places there are coalfields, but the total output of the whole country is very small. One of the large coalfields of England produces hundreds of times the amount of coal produced by the whole of Ireland. **Iron-ore** is found in Antrim between the layers of volcanic rock. The ore seems to have been formed in lakes that occupied the surface of the lava plateau between one volcanic eruption and the next. Many of the Irish **marbles** are of considerable commercial importance. Black marble is worked in Galway and Kilkenny, and red marble in County Cork. The beautiful green marble of Connemara is much used for trinkets and other ornamental work.

Ireland is justly famed for its "**cottage manufactures.**" In most parts of the British Isles the factory operative has supplanted the home worker, but in Ireland several flourishing industries are carried on by the peasants in their own homes. Irish **lace** is highly valued. It is handmade by girls at home, and there are several organisations devoted to raising the already high standard of the lace produced. Although **shirt-making** is now mainly a factory industry, a considerable share of the work is still done in cottages. Cuffs and neck-bands are sewn on, and button-holes are worked by the wives and daughters of the farmers in small holdings in the north of Ireland. Over £50,000 a year is paid to the cottage shirt-workers of Derry, Donegal, and Tyrone. The **woollen goods** of Donegal have a world-wide reputation. The woollen industry of this county is still very largely a purely cottage industry. The girls card and spin the wool from their fathers' sheep. It is woven by a neighbouring weaver, and sold to the local merchant. These home-spun tweeds command a high price. Thousands of the girls are engaged in hand-knitting. The making of "sports coats" is a speciality of Donegal. Recently a few small carpet factories have been started and have proved

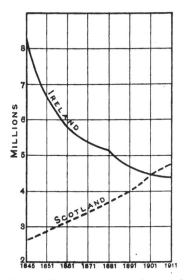

Fig. 72. Graphs showing populations of Ireland and Scotland from 1845 to 1911.

very successful. In the little town of Naas, twenty miles south-west of Dublin, there is rather a remarkable carpet industry. Here rugs and carpets are made for the greatest hotels in London, for the homes of millionaires in America, and for ocean palaces such as the *Titanic* and the *Britannic*. From the designing to the finishing these carpets are purely Irish.

For over half a century the population of Ireland has

been steadily **diminishing**. At the present time the population of Ireland is 4·4 millions, and the population of Scotland is 4·8 millions. At the middle of last century the respective populations were 8¼ millions and 2¾ millions. The decline of the population in Ireland dates from 1846,

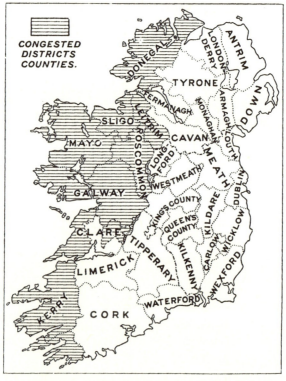

Fig. 73. The Congested Districts of Ireland.

the year of the terrible famine. Owing mainly to emigration the numbers have now shrunk to nearly half their former amount. It was found that parts of the country were unable to support the people settled on them. These parts were called **Congested Districts**, and a Congested Districts

Board was appointed in 1891, which has done an immense amount of good in improving the conditions of the peasants of western Ireland. The Board purchases and resells land to small holders at advantageous rates, and fosters agriculture, forestry, stock-rearing, dairy-farming, poultry-breeding, fishing, spinning, weaving, and other industries, by every means in its power.

CHAPTER XXXIII

ULSTER

THE north-east coast of Ireland at Fair Head is only thirteen miles from the Scottish peninsula of Kintyre. Along this coast the **Antrim Plateau** fronts the sea in lofty cliffs. Inland, the surface of the lava plateau is bare and undulating. A line from Portadown (south of Lough Neagh) to the mouth of Lough Foyle, and another from Portadown to the mouth of Belfast Lough mark the boundaries of the volcanic rocks. **Lough Neagh** is the largest lake in the British Isles, but its shallowness (a maximum depth of 56 feet) indicates a different origin from such lakes as Wastwater and Loch Lomond. It has been formed by a slight subsidence of the volcanic rocks in that neighbourhood. At the **Giant's Causeway** on the northern coast the volcanic rocks have assumed a remarkable appearance. The rock has cracked into regular, vertical columns, generally with six sides, often so perfect that they seem to have been made by human builders. The Giant's Organ, the Giant's Chair, the Giant's Loom, the Giant's Fan—these names indicate the fantastic shapes assumed by the rock. The causeway has been laid bare by the action of the waves, and the sea-caves, reefs, and cliffs of the neighbourhood are all wonderfully picturesque. The pillars of the Giant's Causeway are very like the beautiful columns of Fingal's Cave in Staffa, and, as we have already seen, the volcanic rocks once stretched without a break between Antrim and the Inner Hebrides. The columnar structure has been caused by contraction while the rock was cooling. The same structure can be easily seen in starch, only in this case the contraction is due, not to cooling, but to loss of moisture. The volcanic rocks are seen again in the island of **Rathlin**, where Robert the Bruce spent part of his exile when things were going badly for him in his struggle against the English.

In the north-west of Ulster the bare mountains of **Donegal** attain a height of nearly 2500 feet above sea-level. The scenery is similar to that of the Scottish Highlands, although hardly so grand, for the rocks consist of granites, schists, and gneisses like those of the Highlands. The coasts, too, are fractured and broken with inlets, and fringed with islands. The peasants still speak the old Irish tongue. As we saw in the previous chapter, cottage industries are characteristic of Donegal. The district is particularly noted for its home-spun tweeds and its knitted goods.

The north-east of Ulster is the busiest and most prosperous part of Ireland. A considerable proportion of the inhabitants is not of Irish blood, but descended from immigrants from England and Scotland. The Scottish element preponderates. The **linen** industry is confined mainly to this district. Agriculture is prosperous. The percentage of cultivated land in Armagh and Down is higher than in any other part of Ireland, and the Lagan valley is one of the finest fattening districts for cattle in all Ireland.

Belfast is the largest town, not only in Ulster, but in

ULSTER 203

Ireland. The population in 1911 was 385,000, that is, there are six towns in England larger than Belfast. Its situation, where the Lagan flows into Belfast Lough, is favourable, for it is well placed for trade with England and Scotland. The valley of the Lagan also offers an easy

Fig. 74. The Giant's Causeway.

route to the interior. Belfast is the greatest **manufacturing centre** in Ireland. Indeed it claims to have the largest ship-yard, the largest linen-mill, the largest mineral-water factory, the largest rope-works, and the largest tobacco factory in the world. Whether all these claims can be substantiated or not, it is certain that in all these respects

Belfast carries on a very large trade. The **linen** industry is the most important. Home-grown flax is used, and also material imported from Russia and other countries. Irish linen is unsurpassed for purity and whiteness owing to the sun-bleaching methods employed, and therefore linen manufactured on the Continent is often sent to Belfast to be bleached. Even cotton goods from Manchester are sent for the same purpose. The town is a great **ship-building** centre. Although it cannot produce an aggregate tonnage like that of the whole Clyde district, yet the famous yard of Harland and Wolff has, as a rule, a greater yearly tonnage than any other single yard in Britain. Belfast is, of course, at a disadvantage owing to the paucity of steel works and coal-mines in Ireland, but these disadvantages are counter-balanced by the ease with which coal and iron can be imported from Ayrshire, and the fact that English and Scottish steel-makers are in competition in Ireland, whereas in Great Britain prices are maintained by agreements. The result is that Belfast shipbuilders can buy their steel plates actually at a lower price than the Clyde shipbuilders. In addition to the manufactures already mentioned, **whisky-distilling** and **bacon-curing** are important industries. Most of the trade of Belfast is carried on with Glasgow and Liverpool. A few years ago a university was instituted in Belfast.

CHIEF IMPORTS AND EXPORTS OF BELFAST AND LARNE, 1911

Imports	Value	Exports[1]	Value
Grain and Flour	£2,439,000	Ships	£240,000
Flax	1,895,000	Linen goods and yarn	47,000
Raw cotton	946,000		
Timber	512,000		
Linen yarn	486,000		

[1] The value of the exports of Belfast is very misleading unless we remember that most of the manufactures are sent to Great Britain and therefore are not included under exports.

ULSTER

Londonderry (or Derry, for the prefix was of later origin) is situated on the banks of the River Foyle near the head of Lough Foyle. The principal industry is the manufacture of linen and linen goods, particularly shirt-making. There are large factories in the town itself, and the cottage workers of the surrounding district also share in the trade. There are distilleries which produce not only whisky but a very fine yeast, reputed the best in the country. There are salmon fisheries in Lough Foyle. Derry is the fourth town of Ireland, ranking after Cork in size. It is memorable in history for the great siege in 1689.

Coleraine, at the mouth of the Bann, has industries similar to those of Londonderry. Linen is manufactured, bacon is cured, there are distilleries of whisky, and the salmon fishing in the neighbourhood is excellent. Small ships can come to Coleraine, but large boats use the harbour of **Portrush**, a few miles away. Portrush is the most popular watering-place in Ulster. The cliff scenery in the neighbourhood is magnificent. The black lavas rest directly on chalk beds of the same nature and age as those in the south of England, and the contrast is striking. Portrush is the most convenient centre for visiting the Giant's Causeway. The first electric railway in this country was made from the town to the Causeway. The fine golf-links also attract many visitors.

The principal railway junction of Ulster is **Portadown**, south of Lough Neagh. Lines radiate from Portadown in four directions, one to Belfast, one to Londonderry, one to Dublin, and one south-west to the Central Plain. Belfast is connected with Dublin by the Great Northern Railway. It runs inland up the Lagan valley to Portadown, and then goes due south to the coast, thus avoiding the formidable barrier of the Mourne Mountains. The Belfast and Northern Counties Railway runs north along the coast from Belfast. But soon the hills come right up to the sea, and the line is forced inland in order to avoid the higher parts of the Antrim Plateau. It meets the sea again at Coleraine, and then skirts the low shores of Lough Foyle until Londonderry is reached.

Ulster can be reached from England or Scotland by several fast cross-channel routes. In summer there are daily sailings between Ardrossan and Portrush. Boat trains, of course, connect the former port with Glasgow. The Mull of Kintyre is only a dozen miles from Antrim, but there are no sailings between these points, because these places, being far from any large town, are not served by railways. The **shortest sea-passage** between Great Britain and Ireland is from Stanraer to Larne. It takes two hours to go from port to port. Speedy turbine steamers sail daily between Ardrossan and Belfast. The London and North-Western Railway steamers give a service between Fleetwood and Belfast, and between the latter port and Heysham ply the steamers of the Midland Railway. The same line also runs boats from Barrow to Belfast.

CHAPTER XXXIV

DUBLIN AND THE EAST OF IRELAND

WE saw in the previous chapter that the English and Scottish elements are stronger in Ulster than in any other part of Ireland. This is interesting, because one would expect that the district round Dublin would show alien influences to a higher degree. The human factor has triumphed over purely geographic influences, for the Irish coast between Dublin and Dundalk lies more open to invasion from Great Britain than any other part of the

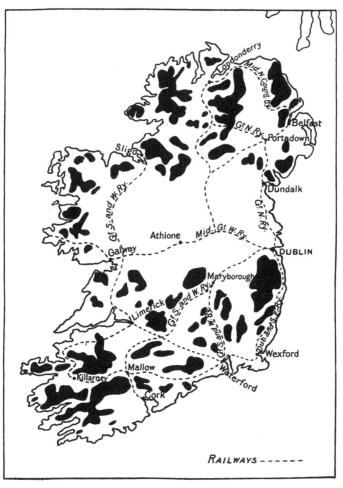

Fig. 75. The Main Railway Lines of Ireland.

country. Between these towns there is a gap in the mountain rim of Ireland sixty miles wide, offering a broad and easy passage to the heart of the country. In addition, this gap lies exactly opposite the western termination of the Midland Gate of England. The central east coast of Ireland has been in close touch with England for a longer period even than Ulster, although settlements have been more permanent in the latter district. For many centuries part of the east coast of Ireland was ruled by the English, although all the rest of the country was quite independent. The strip held by England was called the **Pale**, and stretched roughly from the Wicklow Mountains to the Mourne Mountains.

The **Wicklow Mountains** are the chief highlands of south-eastern Ireland. In Lugnaquilla they attain a height of over 3000 feet above sea-level. Just as the Antrim Plateau was formerly continuous with the Inner Hebrides, so the Wicklow Mountains and the Welsh Highlands originally formed one plateau. The fame of the scenery of the Wicklow Mountains is second only to that of the Killarney district. The celebrated Vale of Ovoca and the beautiful Glendalough can hardly be matched in all Ireland. The district, too, is famous for the ruins of old abbeys, priories, round towers, and other buildings associated with the monks. The towns on the coast are chiefly watering places. Not only is the scenery beautiful, but the climate is dry, for the towns lie in the rain-shadow of the Wicklow Mountains. Of the health resorts **Bray** is the most fashionable. It was originally a tiny fishing village, but it has grown to such an extent in recent years as a watering-place as to merit its name, the " Irish Brighton." **Wicklow**, also, is popular with holiday-makers.

Dublin, at the mouth of the Liffey, is the capital of Ireland, and in size ranks next to Belfast. Its **advantages of position** are exceptional. Being situated in the wide

DUBLIN AND THE EAST OF IRELAND

gap in the mountain rim of Ireland, it has easy access to all parts of the country. This is evident from the way in which the railway lines radiate from Dublin. It is in a good position for trading with Great Britain, and is centrally situated on the east coast, the busiest part of Ireland. Not only is Dublin Bay the best harbour in the neighbourhood but it lies opposite the western end of the Midland Gate of England. At this point, too, the Irish Sea narrows considerably, the distance between Dublin and Holyhead being only sixty miles. Dublin, like London and Glasgow, is a good example of a " bridge town." It is the lowest point at which the Liffey can be conveniently bridged, and from the earliest times there has been a bridge at Dublin.

The position of Dublin is obviously ideal for a capital, particularly for the capital of a country united to Great Britain. Between Dublin and Belfast there is the same difference as between Edinburgh and Glasgow. One town is the political capital, the other the commercial capital. The principal government buildings are in Dublin, and the Viceroy of Ireland has a house in Phoenix Park, Dublin's famous pleasure ground. There are two Colleges, one of which is part of the recently constituted National University, which comprises two other colleges in Cork and Galway, and which is predominantly Catholic. The other is the famous seat of learning, Trinity College, with its memories of Swift, Goldsmith, Burke, and Moore. On the whole, Dublin, although it has slums like other capitals, is a handsome city. Not the least of its attractions are Phoenix Park, in some respects the finest in the British Isles, and the beautiful Dublin Bay.

Although not primarily an industrial centre like Belfast, Dublin has several manufactures of the highest importance. Chief among these is **brewing**. The famous breweries of Guinness are claimed to be the largest in the world. Most of the beer is made from roasted malt, and has therefore a dark colour, and is known as "stout." Two thirds of all the beer made in Ireland, and over 90 per cent. of all the beer exported from Ireland come from Dublin. There are also large whisky distilleries and factories for the making of aerated waters. Poplin, which is a mixture of silk and a cheaper material, often cotton, is made largely in Dublin. The district round the city is a famous region for cattle grazing, and large numbers of cattle are exported from Dublin. For fast mail and passenger traffic, the port of **Kingstown**, six miles down Dublin Bay, is used. Irish time is taken from Dublin Observatory. The longitude of the observatory is 6° 20′ W., whence it follows that when it is noon at Greenwich it is $25\frac{1}{3}$ minutes to twelve at Dublin.

North of Dublin there are no towns of any size until we reach Drogheda. This district is devoted mainly to agriculture and stock-rearing. Dublin County grows a fair amount of **wheat**, indeed more for its size than any other county in Ireland, and Meath is noted for its **cattle**. **Drogheda**, near the mouth of the River Boyne, is interesting chiefly from its historical associations. During the Civil War it was stormed by Cromwell, and the garrison massacred by his orders. At the Revolution the decisive Battle of the Boyne was fought near Drogheda. There are linen factories and breweries in the town. It is also a seaport on a small scale, trading mainly with Liverpool. **Dundalk**, a little farther north, is a town about the same size as Drogheda, and with a similar trade.

A glance at a railway map shows how the railways radiate from Dublin (see Fig. 75). The **Great Southern and Western** line is the most important railway in Ireland. It connects Dublin with the south, the west, and part of the north-west, of Ireland. The map shows that hilly ground is almost continuous from the Slieve Bloom Mountains to

DUBLIN AND THE EAST OF IRELAND

Limerick. Again a series of detached highlands stretches from the Kerry Mountains to the Wicklow Mountains. Between these two highland systems, however, there is a broad band of low ground which is taken advantage of by the main line of the Great Southern and Western Railway. The line goes south-west to Mallow, where it crosses the mountains by a deep gap, and finally turns a little eastwards in order to reach Cork. The gap just south of Mallow is probably the valley of an ancient consequent river beheaded by the Blackwater. A branch of this

Fig. 76. The Mallow Gap. The black areas represent land more than 500 feet above sea-level.

railway runs from Maryborough down the Nore valley to Kilkenny, and then to Waterford.

The **Midland Great Western Railway** goes right across the Central Plain from Dublin to the west coast. At Mullingar it meets a branch from Portadown and the north. It then runs west to Athlone and Galway. The **Dublin and South-Eastern Railway** is pressed close to the coast by the Wicklow Mountains. It turns inland at Wicklow, and runs down the Vale of Ovoca in order to serve the tourist traffic of the Wicklow Mountains. It then runs south to Wexford and Waterford. The **Great Northern Railway** from Dublin to Belfast was mentioned in a former chapter.

Ireland is much better suited than England or Scotland for the development of a good system of **water-ways**. The interior is perfectly flat for long distances, the water-supply is abundant, there are many lakes, and navigation is not impeded by ice in winter. Yet the canal and navigable river systems of Ireland are not particularly good. The population is sparse, there are no large industrial centres in the interior, and much of the farm produce is of so perishable a nature that canal haulage is impossible. The **Grand Canal** is the most important in the country. Its main line goes due west to the Shannon, and then north-west to Ballinasloe. It also brings Dublin into touch by water with Limerick and Waterford. The **Royal Canal** goes north-west from Dublin to Mullingar, and thence to the Shannon. It is owned by the Midland Great Western Railway, but competes unsuccessfully with the railway for the traffic of the Central Plain.

The chief cross-channel routes of this district start from Kingstown or Dublin. The most important route is to Holyhead, where the boats connect with expresses of the London and North-Western Railway. Kingstown can be reached in eight and a half hours from London. This is the chief passenger and mail route. Cattle, agricultural produce, and manufactures go mainly from Dublin to Liverpool.

CHAPTER XXXV

SOUTHERN IRELAND

IN a former chapter we saw that the dominant structural feature of south-west Ireland was the series of ridges and valleys caused by earth movements acting at the end of the Carboniferous epoch. The peculiar character of the **coast-line** of this part of Ireland has been caused by the partial submergence of a land ridged and furrowed by pressure from the south. The map shows that the south-west coast is indented by long bays, the largest of which are Dingle Bay, Kenmare Bay, and Bantry Bay. These

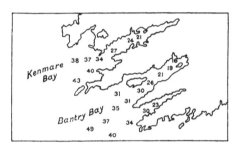

Fig. 77. Rias of south-west Ireland. The figures are soundings in fathoms from the Admiralty Chart.

bays are known as **rias**, a Spanish word, for inlets of this nature are typically developed on the north-west coast of Spain. A ria differs fundamentally from a fiord. From head to mouth a ria gradually widens and becomes deeper, and the floor of the ria is not hollowed out into basins. In these respects the ria contrasts strongly with the fiord. Fig. 77 shows the gradual widening and deepening of Kenmare Bay and Bantry Bay, both typical rias. The characteristics of this coast-line are obviously explained simply by a submergence of a wrinkled area. The bays are drowned valleys, and the peninsulas jutting into the sea are the crests of the ridges. The numerous islands round this coast also point to a submergence of the land. Valentia Island contains a meteorological station which is generally the first to report the oncoming of storms, which in this country usually come from the south-west. One of the most important of the transatlantic cables starts from this Island. Clear Island is the most southerly part of Ireland. It has a lighthouse and a signal station for reporting the movements of ships.

The lake district of **Killarney** is the best known part of south-west Ireland. There are three lakes encircled by mountains. The lakes have been formed by the damming action of moraines. They are studded with islets, and the beauty of the scene is enhanced by several historic ruins. The contrast of lake and mountain makes a lovely picture, but one that could easily be matched in Scotland, Cumberland, or Wales. The peculiar charm of Killarney lies in the marvellous richness and colour of the foliage. The mountains are thickly clothed with trees and flowering plants, the islands are luxuriantly wooded, and the lakes nourish masses of wonderful lilies.

The **richness of the vegetation** of this part of Ireland is, of course, due to the exceptional mildness of the climate. The winters are actually as mild as those of central Italy. The favourable temperature combined with abundant rainfall produces the rich pastures that are so characteristic of southern Ireland. The result is that **dairy-farming** and **stock-rearing** are carried on to a greater extent than in any other part of the British Isles. Practically all the seaports of the south coast are engaged in exporting live stock, meat, and dairy produce. The **butter** of the district is famous, and **eggs** from southern Ireland are sent to practically every town in Great Britain. Even the

SOUTHERN IRELAND

manufactures are those that arise naturally among agricultural communities. **Bacon-curing** factories, many of considerable size, are found in most of the towns. The local barley is used for the distilling of whisky and the brewing of beer and stout. The presence of large numbers of cattle has given rise to the manufacture of leather. In the south-east, where the climate is drier, a considerable amount of grain is grown. **Wheat** and **barley** are grown in the basins of the rivers that flow into Waterford Harbour

Fig. 78. Lower Lake, Killarney.

and Wexford Harbour. More than half the barley grown in Ireland is produced in this district. Because of its fertility the low ground stretching from the Suir past Tipperary to Limerick is known as the Golden Vale.

Cork is the natural capital of south Ireland. Its trade is thoroughly representative of the industries of all the south. It exports more agricultural produce than any other town in the British Isles. County Cork is especially famous for its **butter**, great quantities of which are exported from the county town. There are large **bacon-curing** works and **leather-factories**. The barley of south Ireland is utilised in the **breweries** and **distilleries** of the town. One of the three colleges of the National University is situated in Cork. Cork owes much of its prosperity to its fine harbour, the drowned lower course of the River Lee. Like Dublin, Cork possesses an outport lower down the estuary. This is **Queenstown**, which handles the mails and most of the passenger traffic. Queenstown is the most important calling place for Atlantic liners in Ireland. The mails from America are landed at Queenstown and sent by rail to Kingstown. There they are put on board fast cross-channel steamers for Holyhead, and then carried by express trains to London and other large centres. The fact that Cork is in such close touch with America explains the anomaly that it has a greater foreign trade even than Belfast, a town five times its size. The estuary possesses a Royal Dockyard, and is therefore strongly fortified.

The drowning of the lower courses of the estuaries by the partial subsidence of southern Ireland has converted them into excellent natural harbours. The chief towns are therefore found at the mouths of the principal rivers. Wexford at the mouth of the Slaney, Waterford at the mouth of the Suir, and Youghal at the mouth of the Blackwater are the most important. The termination *ford* in the names of two of these towns does not mean a river crossing, but is really the Scandinavian word *fiord*. The names remind us that the Danes at one time held most of the south Irish coast. Both **Waterford** and **Wexford** were Danish towns. The trade of the coast towns of this part of Ireland is practically the same as that of Cork, although on a smaller scale. All the towns export cattle, butter, eggs, bacon, and pork. Most of the towns have bacon-curing factories, and are also engaged

SOUTHERN IRELAND 217

in brewing and distilling. **Youghal** is associated in history with Sir Walter Raleigh. He was mayor of the town in 1588, and the house he lived in is still preserved. Wexford possesses an outport which, although not yet of the same standing as Kingstown and Queenstown, is rapidly rising in importance. This is **Rosslare**, between which port and Fishguard, just over sixty miles away, ply the fast turbine steamers of the Great Western Railway.

Very few of the inland towns of southern Ireland are of sufficient importance to require mention. The hilly parts of the interior are barren and incapable of supporting many inhabitants. The only well populated parts are the fertile valleys of the Suir and the Nore. **Clonmel** on the Suir and **Kilkenny** on the Nore are the only inland towns of any size in all the south of Ireland. **Cashel** in the Suir basin claims attention for its historic associations. Round the steep hill of Cashel can be seen the ruins of many fine old buildings, including a cathedral, a chapel, and a palace. The town was the ancient capital of the kings of Munster. It is well to remember the high degree of civilisation attained by the Irish when the people of England and Scotland were little more than barbarians.

CHAPTER XXXVI

WESTERN IRELAND AND THE CENTRAL PLAIN

THE most striking feature in the **structure of western Ireland** is the way in which the west coast juts into the Atlantic Ocean between Donegal Bay and Galway Bay. The projecting mass is roughly rectangular in shape and mountainous in character. This part of Ireland stands out in bold relief because of the nature of the rocks composing it. The limestone of the Central Plain extends westwards to Lough Corrib, Lough Mask, and Lough Conn. Farther west the rocks consist of ancient schists and gneisses pierced by granite bosses. These resistant rocks form the mountains of Mayo and Galway. The map shows that a range of hills (the Slieve Gamph and the Ox Mountains) runs north-east from Lough Conn towards the town of Sligo. This is caused by the presence there of a narrow strip of these old rocks, surrounded on all sides by low beds of Carboniferous Limestone. The broken and indented nature of the coast-line of this part of Ireland resembles that of the western Highlands of Scotland, where the rocks are of a similar character. The coast scenery of western Ireland is exceptionally fine. The land fronts the sea in giant cliffs. In Achill Island the breakers of the Atlantic Ocean dash against the foot of precipices that rise sheer for over a thousand feet.

Most of the land of western Ireland is barren, and the population is scanty. Yet there are more people than the land can support, and so all western Ireland comes under the operations of the Congested Districts Board. The inhabitants are mainly crofter-fishermen. In summer and autumn many of them cross to England and Scotland to engage in harvesting, potato-digging, and other temporary farm-work. Sheep are kept on the hill-sides, and cattle in the valleys. The lower Shannon basin is distinctly the most prosperous part of western Ireland. County Clare and County Limerick are both noted cattle-rearing districts. The **Golden Vale** that stretches south-east from Limerick is a rich agricultural region. In this district, too, cottage industries are carried on, of which the most important is lace-making.

The **Shannon** is not only the chief river of western Ireland, but the longest in the British Isles. For much of

AND THE CENTRAL PLAIN

its course it flows slowly over the flat Central Plain. Its banks are low, and at intervals the river widens into a large lake, owing to the solution of the underlying limestone rocks. Below Lough Derg, however, the character of the river alters. It rushes down its channel over a series of rapids, and mountains rise steeply on both sides of the valley. This change in the nature of the Shannon has been attributed to "rejuvenation." Although the main movements of the land have been downwards, a comparatively recent uplift of this part of Ireland has given fresh energy to the lower course of the Shannon, and the valley there has therefore all the characters of a young river. The mouth of the river is an estuary seventy miles long. It is a true ria formed by the drowning of the lower course of the river. Recent investigation has shown that the valley of the Shannon can be traced far out to sea as a submarine trench. It extends as far as the edge of the continental shelf where it descends "through lofty walls of rock into the abyssal floor of the ocean." This indicates that Europe formerly extended westwards at least as far as the present edge of the continental shelf.

Limerick is the only town of any size in western Ireland. It has rather less than forty thousand inhabitants, that is, it is approximately the same size as Chester in England or Kilmarnock in Scotland. It is a typical example of a market town. The surrounding district is rich in crops and cattle, and its trade centres almost exclusively in the county town. It derives additional importance from being situated at the lowest point where the Shannon can be bridged. The industries of the town are those that would naturally arise in the market town of a district devoted to agriculture and stock-rearing. There are flour mills, condensed-milk works, and the largest bacon-curing factories in Ireland. Leather is made, the tanners using local hides and also those imported from South America. The cottage workers of Limerick and the surrounding district are especially famous for their hand-made lace.

Galway has lost much of its former importance. Its trade has dwindled, and one by one its factories have shut down. Sixty years ago the town was nearly double the size it now is. It had formerly a large trade with Spain, but this has now practically disappeared. Galway has many geographical advantages for commerce, but these are all counter-balanced by the fact that its hinterland is poor and unproductive. It has more than once been proposed to make Galway a rival of Queenstown in transatlantic trade. For this purpose the situation of the place is ideal. The route would be shortened, and the fine bay sheltered by the Aran Hills makes a magnificent harbour. Nothing definite, however, has yet been done. Galway is one of the principal stations of the west of Ireland fishing-fleet. It contains one of the colleges of the National University. The Aran Islands, at the mouth of Galway Bay, are of great interest because of the ruins they contain that attest the high civilisation of western Ireland in early times. **Sligo** is the chief fishing port of the north-west. Situated on Sligo Bay among beautiful surroundings it is famed for the number of cromlechs, circles, and other stone remains in the vicinity.

The **Central Plain** of Ireland owes its peculiar character to the nature of the underlying rocks, which are limestones of Carboniferous age. As a rule limestone areas are dry, but the Central Plain is so low-lying that the rocks are water-logged. Wet limestone districts nourish a very rich grass, and this is one reason for the fine grazing grounds of the Irish Plain. One can travel for many miles and see nothing but stretches of green meadow-land varied by lakes and brown peat-bogs. The country seems to consist of an endless succession of small, green fields, divided

Fig. 79. View near Sligo where the Central Plain meets the Western Mountains. In the foreground are peat-stacks and remains of ancient forests. The hill in the background shows the typical appearance of the Millstone Grit.

by wooded turf dykes or by slowly flowing marshy rivers. Few living things are to be seen except oxen, and, in the vicinity of the little farm-houses, a flock of geese, or a donkey, or a few goats. The towns are small and generally at long distances from one another. The **Bog of Allen** which extends from the Shannon to the Boyne is the largest area of bog-land in this country. The bogs were formerly lakes, but are now filled with bog moss. They produce peat, the sole fuel of the Central Plain.

CHAPTER XXXVII

THE ISLE OF MAN AND THE CHANNEL ISLANDS

The **Isle of Man** is almost equidistant from England, Scotland, and Ireland. A circle drawn with its centre at Douglas and a radius of 42 miles would just graze the coasts of Cumberland and Down, and would cut off the tips of the two peninsulas of Wigtown. The Isle of Man, however, is much more closely connected with England than with Ireland, for between it and the latter country there is a submarine trench more than fifty fathoms deep, while between Man and England the sea is less than twenty fathoms deep. The island is very low in the north, but in the centre rises to over two thousand feet in **Snaefell**. The rocks of Man are very ancient, and probably were once continuous with those of Cumberland, which they resemble in many respects although somewhat older. They consist of slates and grits which in places have been so folded and crumpled by earth movements that they have been crushed into a conglomerate. Along the coast they form fine cliffs. At Peel and at Castletown there are small patches of

THE ISLE OF MAN

Carboniferous Limestone, and the northern part of the island is covered with sand and boulder-clay.

The Isle of Man contains good pasture land, and, therefore, many cattle and sheep are kept. The fisheries round the coast are very important, and give employment to thousands of Manxmen. Some minerals are obtained of which the chief is lead. The lead mines at **Laxey** are among the largest in the British Isles. A great many people make a living by catering for summer visitors, for the Isle of Man is one of the most popular holiday resorts in this country. All tastes may be satisfied. The scenery is beautiful, particularly the cliff scenery of the southern coast. The climate is dry and sunny in summer, and mild in winter. At **Douglas** and **Ramsey**, the largest towns, there are crowds, bands, and gaiety, while the smaller places appeal to the geologist, the archaeologist, and the lover of beautiful scenery.

The Isle of Man has a Parliament and laws of its own, and is not affected by the laws made at Westminster unless specially named. There is, however, a Governor appointed by the Crown. The Manx language belongs to the Celtic group, and is very similar to old Irish. It is rapidly falling into disuse. The people are mainly of Celtic blood with a strong Scandinavian element. Indeed for three hundred years the Isle of Man belonged to Norway. As we have seen elsewhere in Britain, the place-names bear witness to the history of the Island. Some of the **Manx names** are plainly Celtic, like *Ballabeg*, *River Dhoo*, and *Bride*, while others are just as clearly of Norse origin as *Snaefell*, *Laxey*, and many names ending in *by*.

In most respects the **Channel Islands** are more akin to France than they are to England. **Alderney** is nearest both to England and to France, but while it is fifty miles from the former country, it is only ten miles from the latter. The rock structure of the islands, too, allies them to France.

They are granite bosses and resemble the granites of the neighbouring Cotentin Peninsula. Many of the people still speak a dialect of French, and French is still the official language of the Islands. The islands belonged to Normandy when Duke William conquered England, and ever since have remained part of the British Isles. As in the Isle of Man, there is a separate legislature, and British laws must specially name the Channel Islands if they are to apply to these islands.

The climate is exceptionally mild and pleasant. The islands are favourite places of residence for invalids in winter, for frost and snow are seldom found. The chief branch of agriculture is the growing of early vegetables, flowers, and fruit for the English market. The principal islands are **Jersey**, **Alderney**, and **Guernsey**. Each island is noted for a special breed of cattle. There are regular sailings between the islands and Southampton.

INDEX

Abbotsford 79
Aberdeen 74
Aberystwith 183
Advantages of position of British Isles 2
Agriculture, contrast between W. and E. of Scotland 78, 89
  of British Isles 30
  of Ireland 196
Airdrie 85
Aire 107
  Gap 160, 187
Alderney 223, 224
Aldershot 120
Allen, Bog of 192, 222
Alloa 94
Altitude, effect on agriculture 30
Animals of England and Continent 7
Anticyclone 18
Antrim Plateau 201
Arbroath 94
Arctis 51, 60
Ardrossan 89
Armorican Mountains 194
Arran 88
Atlantic Drift 28
Avon, piracy by 53
Aylesbury 121
Ayr 89
Ayrshire Plain 89
Azores, high pressure centre of 18

Bandon 194
Barmouth 183
Barrhead Gap 196
Barrow 194
Barrow-in-Furness 167
Bath 139
Bathgate 92
Beachy Head 123
Bedford 152
Beheaded streams 53, 108, 117, 194, 211

Belfast 202
Belfast and Northern Counties Railway 205
Birkenhead 166
Birmingham 143
Blackburn 162
Black Country 143
Blackpool 166
Blackwater 194, 216
Bodmin Moors 129
Bolton 162
Bore of Severn 135
Bradford 170
Bray 208
Brecknock Beacons 176
Brighton 127
Bristol 138
  Channel 6
  Coalfield 139
British Isles joined to Continent 7
Broads 147
Broxburn 93
Buchan, Plain of 74
Building stone 44
Build of British Isles 9
  of England 99
  of Ireland 190
  of Scotland 44
Burnley 162
Burton-on-Trent 144
Buxton 158

Cairngorm Mountains 72
Caledonian Canal 72
  Railway 80, 95
Camborne 133
Cambridge 149
Campbeltown 73
Campsie Fells 82
Canals of Ireland 212
Canterbury 126
Capes, formation of 123
Cardiff 181
Carlisle 168, 186, 187

M.   15

INDEX

Carnarvon Castle 180
Cashel 217
Central Lowlands 10, 82
    railways of 94
    rocks of 46
    structure of 83
    Plain of Ireland 220
Chalk ridge 103
    rocks 15
Channel Islands 223
    climate of 224
Charnwood Forest 102
Chatham 122
Cheddar Gorges 139
Cheviot Hills 77
Chiltern Hills 15, 104
Cleveland Hills 15, 103, 169
Climate and character 2
    of British Isles 15
    of Ireland 195
Climatic control of cotton industry 164
Clonmel 217
Clyde 53
    shipbuilding 86
    Valley 84
Coal 36
    exports of 40
Coalfields and large towns 176
    of the British Isles 38
Coal Measures 14
    output from U.K., U.S.A., and Germany 39
    reserves of 40
Coast erosion 147
    line of Southern Ireland 213
    line, origin of features of 8, 59
Coatbridge 85
Coires, origin of 69
Coleraine 205
Congested Districts 200
Consequent rivers 52
Continental pressure centre 18
    Shelf 4
Contours of sea-floor 4
Contrast between N.W. and S.E. of the British Isles 10
Conway Castle 180
Copper 43
Cork 215
Cottage manufactures of Ireland 198
Cotteswold Hills 15, 103
Cotton, imports of 162
    industry, climatic control of 164
Counties of England 188
    of Scotland 96

Counties, origin of 96, 188
Coventry 144
Cowes 128
Crewe 145, 186
Cross Fell 157
Cuchullin Hills 50, 64
Cumberland Coalfield 168
Cyclone 16

Dairy-farming 34
Dalbeattie 78
Darlington 174
Dartmoor 101, 129, 130
Deer-forests 73
Derby 145, 187
Derg, Lough 219
Derwent 107
Derwentwater 154
Devonport 133
Devon, seamen of 133
Dingle Bay 213
Dissected plateau 12
    of Scottish Highlands 70
Dogger Bank 6
Don, beheading of 53
Donegal 202
Douglas 223
Dover 127
Dredgings from North Sea 7
Droitwich 135
Drowned rivers 59, 132, 214, 216, 219
Dublin 208
Dublin and South Eastern Railway 211
Dudley 143
Dumbarton 90
    Rock 66
Dumfries 79
Dundee 94
Dunfermline 93
Dunoon 88
Dunstable 118
Durham 174

East Anglian Heights 146
Eastbourne 127
Eastern England 146
    agriculture of 148
    fisheries of 148
    rocks of 146
    Ireland 206
Eden 110
Edinburgh 90
    Rock 66
Electric power in Highlands 73

INDEX 227

England, build of 99
  joined to Continent 7
  rivers of 105
English counties, origin of 188
Equinoctial gales 21
Escarpments 102
Eton 120
Exeter 10, 133
Exmoor 101, 130
Extension west of British Isles 8

Falkirk 92
Falls of Clyde 85
Faults 45
Fens 147
Fife Coalfield 93
Fifeshire 93
Fiords of Scotland 58
Firth of Clyde 88
  of Forth 92
Fisheries of British Isles 35
  of Ireland 197
Fishguard 183
Fishing grounds 6
  towns 36
Fish, value in 1910 36
Flax 33
Fleetwood 166
Folded mountains 12
Folkestone 127
Forth 56
Fort William 74
Fracturing of British Isles 8, 11, 59, 60
Fruit 34

Galashiels 79
Galty Mountains 193
Galway 220
Gaps in Chilterns 108
  in Downs 125, 126
Gap towns of England 126
  of Scottish Lowlands 90
Gateshead 174
Giants' Causeway 201
Glaciers 65, 66
Glasgow 84
  and South Western Railway 80, 96
Glencoe 74
Glendalough 192, 208
Glen More 10
Gloucester 136
Golden Vale 215, 218
Gorges of Cheddar 139
Goring Gap 108
Grain crops of British Isles 32
Grampian Highlands 10, 72

Grand Canal 212
Grangemouth 92
Granite bosses 65
  of S.W. England 129
Great Central Railway 187
  Eastern Railway 188
  Northern Railway 187
  Northern Railway of Ireland 205, 212
  North of Scotland Railway 75
  Southern and Western Railway 210
  Western Railway 187
Greenock 88, 122
Grimsby 150
Guernsey 224
Guildford 126
Gulf of Warmth 28, 29
  Stream Drift 28

Halifax 170
Hamilton 85
Hampshire Basin 105, 123
Harrogate 172
Harrow 120
Harwich 149
Hawick 79
Hebridean Gneiss 62
Hebrides 6
  Inner 60, 64
  Outer 59, 62
  rocks of 49, 62, 64
  scenery of 62, 64
Helvellyn 154
Hereford 137
Heysham 166
Highland Boundary Fault 45
  Railway 75
Highlands, eastern coastal strip of 49
  of Scotland 70
  rocks of 45
High Wycombe 117
Holyhead 183
Hops 34
Huddersfield 170
Hull 171

Ice, effect on scenery 67
Icelandic low pressure centre 18
Ilfracombe 133
Imports of wheat 32
Industrial Inertia 156, 182
Industries of Ireland 198
Ingleborough 157
Inner Hebrides 60, 64
Inverness 73

Ipswich 150
Ireland, agriculture of 196
   canals of 212
   Central Plain of 220
   climate of 195
   Congested Districts of 200
   fisheries of 197
   industries of 198
   minerals of 198
   mountains of 12
   population of 199
   railways of 205, 210
   rias of 213
   rivers of 194
   rocks of 191
   steamer routes to 206, 210, 212, 216, 217
   structure of 190
   vegetation 214
Irish Sea 6
Iron 41
Iron-ore, deposits of 42
   imports of 42
Islands of Scotland 57
   origin of 60
Islay 64
Isle of Man 6, 222
   of Wight 128
Isotherms, abnormal direction of 28
   for January and July 24, 27

Jersey 224
Jura 64

Kaolin 130
Kenmare Bay 213
Kerry Mountains 193
Keswick 156
Kidderminster 136
Kilbarchan Hills 82
Kilkenny 217
Killarney 193, 214
   origin of lakes of 68
Kilmarnock 89
Kingstown 210
Kirkcaldy 93
Kirkwall 62

Lake District 100
   compared with Wales 178
   Mountains 10
   origin of lakes of 154
   rocks of 100, 153
   scenery of 154
   waterfalls of 156
Lakes of glacial origin 68

Lakes of Scotland 57
Lammermoor Hills 77
Lanark 85
Lanarkshire Coalfield 85
Lancashire Coalfield 161
   cotton industry of 162
Lancaster 167, 186
Land hemisphere, centre of 1
Large towns and coalfields 176
Lead 42
Lee 194, 216
Leeds 170, 187
Leicester 144, 145, 187
Leith 92
Lerwick 62
Limestone bands 15
   regions 158, 220
   ridge 11
Lincoln 151
   Heights 146
Lincolnshire, place-names of 152
Lincoln Wolds 104, 146
Linlithgow 93
Liverpool 162, 165
Llandudno 183
Loch Fyne 59
   Katrine 58
   Lomond 58
   Long 59
   Morar 58
Lochs of Scotland 57
Lodore Falls 156
London 111
   and North-Western Railway 185
   and South-Western Railway 187
   Basin 104
   Brighton, and South Coast Railway 188
Londonderry 205
Lothians 92
Lough Neagh 201
Lowestoft 149
Lowlands, Central 82
   North-Eastern 75
Luton 118

Malvern Hills 102
Man, industries of 223
   Isle of 222
   rocks of 222
Manchester 162
   as a seaport 164
   Ship Canal 162, 163
Manx language 223
Margate 127
Maryport 168

INDEX

Matlock 158
Measurements of length and area 9
Melrose Abbey 79
Mendip Hills 139
Merrick 77
Merthyr Tydfil 182
Methil 94
Middlesbrough 176
Midland counties 190
   Gate 102, 140, 208
   Great Western Railway 211
   Railway 160, 186
Midlands 140
   rocks of 141
Milford Haven 183
Minch 50
Minerals of British Isles 36
   of Ireland 198
Montrose 94
Moorfoot Hills 77
Motherwell 85
Mountains, main masses of 10
Mourne Mountains 191, 193
Mull 50, 64

Needles 124
Newcastle 10, 174
Newhaven 127
Nith 54
Nore 217
Norfolk Broads 147
Northampton 144
North British Railway 80, 95
   Downs 15, 104, 123
North-East England 169
   rocks of 169
North-Eastern Lowlands 75
   Railway 187
North Sea 6
   Staffs. Coalfield 142
   Shields 174
Northumberland and Durham Coalfield 173
North-West England 161
   rocks of 161
   Highlands 10
Norwegian Depression 5
Norwich 150
Nottingham, 145, 187

Oban 74
Obsequent rivers 54
Ochil Hills 82, 83
Oil-shale industry 92
Oldham 162
Oolitic limestone ridge 103

Orkney Islands 60, 62
Ouse 106, 189
Outer Hebrides 59, 62
Ovoca, Vale of 192, 208
Oxford 118

Paisley 88
Pale 208
Peak District 157, 158
Peebles 79
Pennine Uplands 10, 100, 156
   rocks of 100, 157
   routes across 159
   scenery of 157
Pentland Hills 82
Penyghent 157
Penzance 133
Perth 90
Peterborough 152
Place-names of Lincolnshire 152
   of Man 223
   of Wales 183
   of Weald 104
Plain of York 169
Plymouth 133
Plynlimmon 176
Population of Ireland 199
Portadown 205
Port Glasgow 86
Portree 64
Portrush 205
Portsmouth 128
Potteries 142
Preston 162, 186
Pressure centres 18
   charts 16, 17

Queenstown 216

Race influence 2
Railway centres and strategic towns 90, 126, 136, 168
Railways of Central Lowlands 94
   of England 184
   of Ireland 205, 210
   of Pennine Uplands 159
   of Southern Uplands 80
   of Ulster 205
Rainfall of British Isles 23
   Map 25
   of Ben Nevis 23
   of Fort William 23
   throughout the year 24
Ramsey 223
Ramsgate 127
Rathlin 202

Reading 119
Red Plain 101
Renfrew 86
Rias 132, 213, 219
Ribble 107
Ridges 11, 12
River capture 52, 110, 194
    piracy 52, 110, 194
Rivers of England 105
    of Ireland 194
    of Scotland 51
Rocks, age of 14
    and relief 14
    of Central Lowlands 46
    of Highlands 45
    of Inner Hebrides 49
    of Ireland 191
    of Lake District 153
    of Pennine Uplands 157
    of Southern Uplands 47
    of Wales 178
Rock structure 13
Rosslare 217
Rothesay 88
Routes across Pennines 159
Royal Canal 212
Rugby 145, 186
Runcorn 166

St Andrews 94
St George's Channel 6
St Helens 166
St Ives 133
St Paul's 114
Salford 162
Salisbury 125
Salt 43
Scafell 154
Scarborough 172
Scarped lands 102
Scenery, ice-sculptured 68
    of Donegal 202
    of Hebrides 62, 64
    of Lake District 154
    of limestone regions 158
    of Pennines 157
    of Wales 177
    types of 47
Scilly Isles 129, 132
Scotland, build of 44
    compared with Wales 179
    counties of 96
    lakes, lochs, and islands of 57
    rivers of 51
Scottish Highlands 10
    industries of 73

Seamen of Devon 133
Seas of Britain 4
Severn 134
    bore of 135
    tunnel 139
Shallowness of seas 6
Shannon 218
Shap Summit 186
Sheep rearing 34
Sheerness 122
Sheffield 171, 187
Shetland Islands 60
Shipbuilding on Clyde 86
    on Tyne 173
Shrewsbury 136
Sidlaw Hills 82
Situation of British Isles 1
Skiddaw 154
Skye 50, 64
Slaney 194, 216
Slates 44
Slieve Bloom Mountains 192
Sligo 220
Snaefell 222
Snowdon 176
Southampton 127
South Downs 15, 104, 123
    Eastern and Chatham Railway 188
Southern England 122
    Ireland 213
        industries of 214
        rivers of 194
    Uplands 10, 77
        Boundary Fault 45
        historical associations of 80
        railways of 80
        rocks of 47
        scenery of 77
South Shields 174
    Staffs. Coalfield 142
    Wales Coalfield 180
South-West England, climate of 132
    Ireland, rocks of 193
South-Western Peninsula 129
Spey 55
Stafford 145, 186
Stirling 90
    Rock 66
Stock rearing 34
Stockton 174
Stoke 142
Stonehenge 124
Stonehouse 133
Storms, frequency of throughout year 22
Stornoway 64

INDEX

Stourbridge 136
Stranraer to Larne 80
Strategic towns and railway centres 90, 126, 136, 168
Stratford-on-Avon 144
Strathmore 83
Structure of Ireland 190
Subsequent rivers 52
Suir 194, 216, 217
Sunderland 174
Sunshine 29
Swansea 181

Tay 55
Tees as a county boundary 189
Temperature, effect of Gulf Stream on 28
    of Britain compared with Labrador 26
    of British Isles 24
    of east coast compared with west coast 26
Thames 108, 117, 188
    valley 117
Thurso 75
Tidal wave 6
Tin 42
Tobermory 64
Torquay 133
Tower of London 114
Trent 107
Trossachs 58
Truro 133
Tweed 56
Tyne Gap 159
Tynemouth 174
Tyne shipbuilding 173

Ullswater 155
Ulster 201
    industries of 202
    railways of 205

Vale of Ovoca 192, 208
Vegetation of British Isles 30
Volcanic necks 66
Volcanoes 65
    of Inner Hebrides 49

Wales 100, 176
    compared with Lake District 178
        with Scotland 179
    ice-erosion in 177
    industries of 182
    rainfall of 178
    rocks of 100, 178

Wales, scenery of 177
Walsall 143
Warwick 144
Wastwater 155
Waterford 216
Watershed 11
Watling Street 112
Weald 104
    rivers of 109
Weather of British Isles 15
Welsh Mountains 10
    place-names 183
Western Ireland, industries of 218
    rocks of 217
West Highland Railway 76
Westminster Abbey 112
West of England woollen industry 140
Wexford 216
Wheat, imports of 32
Whernside 157
Whitby 172
Whitehaven 168
Wick 75
Wicklow 208
Wicklow Mountains 192, 193, 208, 211
Wigan 186
Winchester 125
Windermere 154, 156
Wind roses 20
Winds, effect on trees 21
    equinoctial gales 21
    in spring 20
    in summer 19
    in winter 19
    of British Isles 18
Windsor 120
Winds throughout year 20
Wishaw 85
Witham 107
Wolverhampton 143
Wool, imports of 170
Woolwich 122
Worcester 136
World-position of British Isles 1
Wye valley 138

Yarmouth 149
York 171
Yorks., Derby, and Notts. Coalfield 170
Yorkshire Wolds 104, 169
    woollen industry of 170
Youghal 217

Zinc 43